Graduate Texts in Mathematics 10

M. M. Cohen

A Course in Simple-Homotopy Theory

Springer-Verlag New York · Heidelberg · Berlin

Marshall M. Cohen
Associate Professor of Mathematics, Cornell University, Ithaca

AMS Subject Classification (1970)
57 C10

Library of Congress Catalog Card Number 72–93439.

Printed in the United State of America.

9 8 7 6 5 4 3 2

ISBN 0–387–90056–X Springer-Verlag New York Heidelberg Berlin
ISBN 3–540–90056–X Springer-Verlag Berlin Heidelberg New York

To
Avis

PREFACE

This book grew out of courses which I taught at Cornell University and the University of Warwick during 1969 and 1970. I wrote it because of a strong belief that there should be readily available a semi-historical and geometrically motivated exposition of J. H. C. Whitehead's beautiful theory of simple-homotopy types; that the best way to understand this theory is to know how and why it was built. This belief is buttressed by the fact that the major uses of, and advances in, the theory in recent times—for example, the *s*-cobordism theorem (discussed in §25), the use of the theory in surgery, its extension to non-compact complexes (discussed at the end of §6) and the proof of topological invariance (given in the Appendix)—have come from just such an understanding.

A second reason for writing the book is pedagogical. This is an excellent subject for a topology student to "grow up" on. The interplay between geometry and algebra in topology, each enriching the other, is beautifully illustrated in simple-homotopy theory. The subject is accessible (as in the courses mentioned at the outset) to students who have had a good one-semester course in algebraic topology. I have tried to write proofs which meet the needs of such students. (When a proof was omitted and left as an exercise, it was done with the welfare of the student in mind. He should do such exercises zealously.)

There is some new material here[1]—for example, the completely geometric definition of the Whitehead group of a complex in §6, the observations on the counting of simple-homotopy types in §24, and the direct proof of the equivalence of Milnor's definition of torsion with the classical definition, given in §16. But my debt to previous works on the subject is very great. I refer to [Kervaire-Maumary-deRham], [Milnor 1] and above all [J. H. C. Whitehead 1,2,3,4]. The reader should turn to these sources for more material, alternate viewpoints, etc.

I am indebted to Doug Anderson and Paul Olum for many enlightening discussions, and to Roger Livesay and Stagg Newman for their eagle-eyed reading of the original manuscript. Also I would like to express my appreciation to Arletta Havlik, Esther Monroe, Catherine Stevens and Dolores Pendell for their competence and patience in typing the manuscript.

My research in simple-homotopy theory was partly supported by grants from the National Science Foundation and the Science Research Council of Great Britain. I and my wife and my children are grateful to them.

Cornell University
Ithaca, New York
February, 1972

Marshall M. Cohen

[1] Discovered by me and, in most instances, also by several others. References will be given in the text.

TABLE OF CONTENTS

A Course in Simple-Homotopy Theory

Chapter I

Introduction

This chapter describes the setting which the book assumes and the goal which it hopes to achieve.

The setting consists of the basic facts about homotopy equivalence and CW complexes. In §1 and §3 we shall give definitions and state such facts, usually without formal proof but with references supplied.

The goal is to understand homotopy theory geometrically. In §2 we describe how we shall attempt to formulate homotopy theory in a particularly simple way. In the end (many pages hence) this attempt fails, but the theory which has been created in the meantime turns out to be rich and powerful in its own right. It is called simple-homotopy theory.

§1. Homotopy equivalence and deformation retraction

We denote the unit interval [0,1] by I. If X is a space, 1_X is the identity function on X.

If f and g are maps (i.e., continuous functions) from X to Y then *f is homotopic to g*, written $f \simeq g$, if there is a map $F: X \times I \to Y$ such that $F(x,0) = f(x)$ and $F(x,1) = g(x)$, for all $x \in X$.

$f: X \to Y$ is a *homotopy equivalence* if there exists $g: Y \to X$ such that $gf \simeq 1_X$ and $fg \simeq 1_Y$. We write $X \simeq Y$ if X and Y are homotopy equivalent.

A particularly nice sort of homotopy equivalence is a *strong deformation retraction*. If $X \subset Y$ then $D: Y \to X$ is a strong deformation retraction if there is a map $F: Y \times I \to Y$ such that

(1) $F_0 = 1_Y$
(2) $F_t(x) = x$ for all $(x,t) \in X \times I$
(3) $F_1(y) = D(y)$ for all $y \in Y$.

(Here $F_t: Y \to Y$ is defined by $F_t(y) = F(y,t)$.) One checks easily that D is a homotopy equivalence, the homotopy inverse of which is the inclusion map $i: X \subset Y$. We write $Y \searrow X$ if there is a strong deformation retraction from Y to X.

If $f: X \to Y$ is a map then *the mapping cylinder M_f* is gotten by taking the disjoint union of $X \times I$ and Y (denoted $(X \times I) \oplus Y$) and identifying $(x,1)$ with $f(x)$. Thus

$$M_f = \frac{(X \times I) \oplus Y}{(x,1) = f(x)}.$$

The identification map $(X \times I) \oplus Y \to M_f$ is always denoted by q. Since

$q|X \times [0,1)$ and $q|Y$ are embeddings, we usually write $q(X \times 0) = X$ and $q(Y) = Y$ when no confusion can occur. We also write $q(z) = [z]$ if $z \in (X \times I) \oplus Y$.

(1.1) *If $f: X \to Y$ then the map $p: M_f \to Y$, given by*

$$p[x,t] = [x,1] = [f(x)], \qquad t < 1$$

$$p[y] = [y] \qquad y \in Y$$

is a strong deformation retraction.

The proof consists of "sliding along the rays of M_f." (See [Hu, p. 18] for details.) □

(1.2) *Suppose that $f: X \to Y$ is a map. Let $i: X \to M_f$ be the inclusion map. Then*

(a) *The following is a commutative diagram*

$$\begin{array}{ccc} X & \xrightarrow{\ i\ } & M_f \\ {\scriptstyle f}\downarrow & \swarrow{\scriptstyle p} & \\ Y & & \end{array}$$

(b) *i is a homotopy equivalence iff f is a homotopy equivalence.*

Part (a) is clear and (b) follows from this and (1.1). □

§2. Whitehead's combinatorial approach to homotopy theory

Unfortunately, when given two spaces it is very hard to decide whether they are homotopy equivalent. For example, consider the 2-dimensional complex H—"the house with two rooms"—pictured at the top of page 3. H is built by starting with the wall $S^1 \times I$, adding the roof and ground floor (each a 2-disk with the interior of a tangent 2-disk removed), adding a middle floor (a 2-disk with the interiors of two 2-disks removed) and finally sewing in the cylindrical walls A and B. As indicated by the arrows, one enters the lower room from above and the upper room from below. Although there seems to be no way to start contracting it, this space is actually contractible (homotopy equivalent to a point). It would be nice if homotopy theory could tell us why in very simple terms.

In the 1930's one view of how topology ought to develop was as *combinatorial topology*. The homeomorphism classification of finite simplicial complexes had been attacked (most significantly in [Alexander]) by introducing elementary changes or "moves", two complexes K and L being "combinatorially equivalent" if one could get from K to L in a finite sequence of such moves. It is not surprising that, in trying to understand homotopy equivalence, J. H. C. Whitehead—in his epic paper, "Simplicial spaces, nucleii and

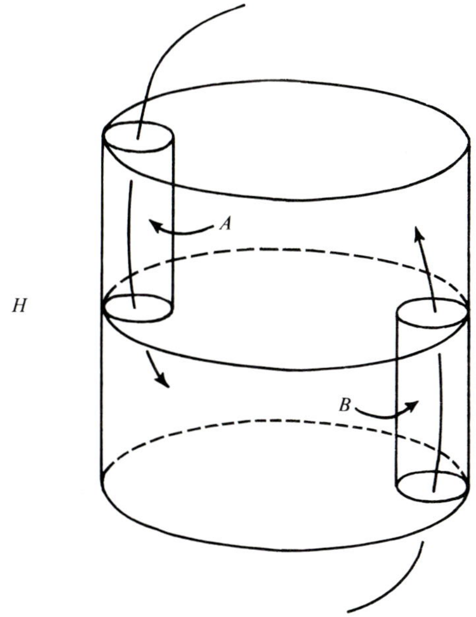

m-groups"—proceeded in the same spirit. We now describe the notions which he introduced.

If K and L are finite simplicial complexes we say that there is an *elementary simplicial collapse* from K to L if L is a subcomplex of K and $K = L \cup aA$ where a is a vertex of K, A and aA are simplexes of K, and $aA \cap L = a\dot{A}$. Schematically,

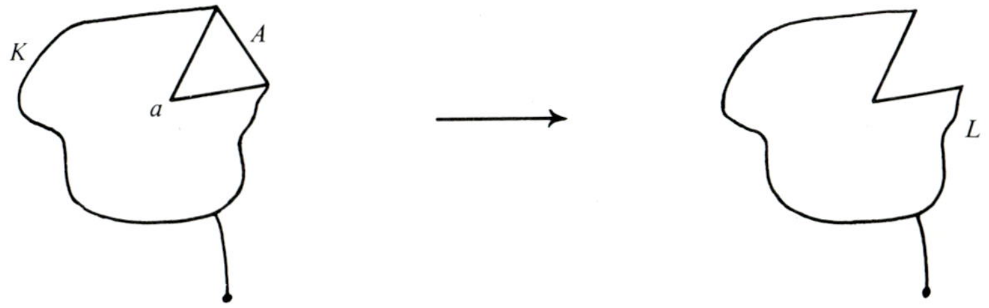

We say that K *collapses simplicially to* L—written $K \overset{s}{\searrow} L$—if there is a finite sequence of elementary simplicial collapses $K = K_0 \to K_1 \to \ldots \to K_q = L$. For example, any simplicial cone collapses simplicially to a point.

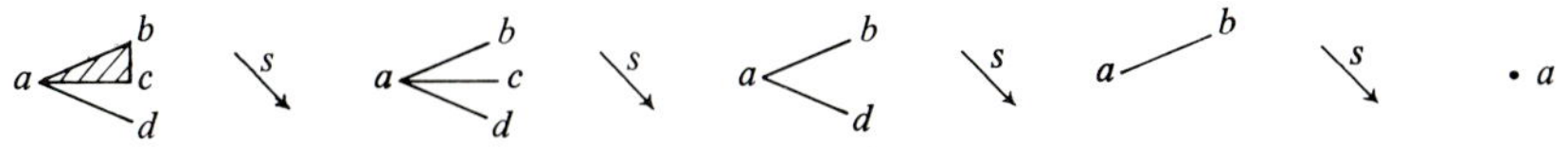

If $K \overset{s}{\searrow} L$ we also write $L \overset{s}{\nearrow} K$ and say that L *expands simplicially to* K. We say that K and L have the same *simple-homotopy type*[2] if there is a finite

[2] This is modern language. Whitehead originally said "they have the same *nucleus*."

sequence $K = K_0 \to K_1 \to \ldots \to K_q = L$ where each arrow represents a simplicial expansion or a simplicial collapse.

Since an elementary simplicial collapse easily determines a strong deformation retraction (unique up to homotopy) it follows that, if K and L have the same simple-homotopy type, they must have the same homotopy type. WHITEHEAD asked

> *If two finite simplicial complexes have the same homotopy type, do they necessarily have the same simple-homotopy type?*

Despite the apparent restrictiveness of expanding and collapsing, it is quite conceivable that the answer to this question might be yes. To illustrate this and to show that simple-homotopy type is a useful notion, let us return to the house with two rooms.

Think of H as being triangulated as a subcomplex of the solid cylinder $D^2 \times I$ where $D^2 \times I$ is triangulated so that $D^2 \times I \overset{s}{\searrow} D^2 \times 0 \overset{s}{\searrow} *$ (= point). Now, if the solid cylinder were made of ideally soft clay, it is clear that the reader could take his finger, push down through cylinder A, enter the solid lower half of $D^2 \times I$ and, pushing the clay up against the walls, ceiling and floor, clear out the lower room in H. Symmetrically he could then push up the solid cylinder B, enter the solid upper half and clear it out. Having done this, only the shell H would remain. Thus we can see (although writing a rigorous proof would be unpleasant) that

$$* \overset{s}{\nearrow} (D^2 \times I) \overset{s}{\searrow} H.$$

Hence H has the same simple-homotopy type as a point and, a fortiori, H is contractible.

So we shall study the concept of simple-homotopy type, because it looks like a rich tool in its own right and because, lurking in the background, there is the thought that it may be identical with homotopy type.

In setting out it is useful to make one technical change. Simplicial complexes are much too hard to deal with in this context. WHITEHEAD's early papers [J. H. C. WHITEHEAD 1, 2] are a marvel in that, besides the central concepts introduced, he overcame an enormous number of difficult technical problems related to the simplicial category. These technical difficulties later led him to create CW complexes [J. H. C. WHITEHEAD 3] and it is in terms of these that he brought his theory to fruition in [J. H. C. WHITEHEAD 4]. In the next section we summarize the basic facts about CW complexes. In Chapter II the expanding and collapsing operations are defined in the CW category and it is in this category that we set to work.

§3. CW complexes

In this section we set the terminology and develop the theorems which will be used in the sequel. Because of the excellent treatments of CW complexes which exist (especially [SCHUBERT] and [G. W. WHITEHEAD]) proofs of standard facts which will be used in a standard fashion are sometimes omitted. The reader is advised to read this section through (3.6) now and to use the rest of the section for reference purposes as the need arises.

A *CW complex* K is a Hausdorff space along with a family, $\{e_\alpha\}$, of open topological cells of various dimensions such that—letting $K^j = \bigcup \{e_\alpha | \dim e_\alpha \leq j\}$—the following conditions are satisfied:

CW 1: $K = \bigcup_\alpha e_\alpha$, and $e_\alpha \cap e_\beta = \emptyset$ whenever $\alpha \neq \beta$.

CW 2: For each cell e_α there is a map $\varphi_\alpha : Q^n \to K$, where Q^n is a topological ball (homeomorph of $I^n = [0,1]^n$) of dimension $n = \dim e_\alpha$, such that

(a) $\varphi_\alpha | \mathring{Q}^n$ is a homeomorphism onto e_α.
(b) $\varphi_\alpha(\partial Q^n) \subset K^{n-1}$

CW 3: Each $\bar{e}_{\alpha 0}$ is contained in the union of finitely many e_α.

CW 4: A set $A \subset K$ is closed in K iff $A \cap \bar{e}_\alpha$ is closed in $\bar{e}_\alpha$ for all e_α.

Notice that, when K has only finitely many cells, CW 3 and CW 4 are automatically satisfied.

A map $\varphi : Q^n \to K$, as in CW 2, is called a *characteristic map*. Clearly such a map φ gives rise to a characteristic map $\varphi' : I^n \to K$, simply by setting $\varphi' = \varphi h$ for some homeomorphism $h : I^n \to Q^n$. Thus we usually restrict our attention to characteristic maps with domain I^n, although it would be inconvenient to do so exclusively. Another popular choice of domain is the n-ball $J^n = \text{Closure}(\partial I^{n+1} - I^n)$.

If $\varphi : Q^n \to K$ is a characteristic map for the cell e then $\varphi | \partial Q^n$ is called *an attaching map* for e.

A *subcomplex* of a CW complex K is a subset L along with a subfamily $\{e_\beta\}$ of the cells of K such that $L = \bigcup e_\beta$ and each $\bar{e}_\beta$ is contained in L. It turns out then that L is a closed subset of K and that (with the relative topology) L and the family $\{e_\beta\}$ constitute a CW complex. If L is a subcomplex of K we write $L < K$ and call (K,L) a *CW pair*. If e is a cell of K which does not lie in (and hence does not meet) L we write $e \in K - L$.

Two CW complexes K and L are *isomorphic* (denoted $K \cong L$) if there exists a homeomorphism h of K onto L such that the image of every cell of K is a cell of L. In these circumstances h is called a CW isomorphism. Clearly h^{-1} is also a CW isomorphism.

An important property of CW pairs is the **homotopy extension property**:

(3.1) *Suppose that* $L < K$. *Given a map* $f : K \to X$ (X any space) *and a homotopy* $f_t : L \to X$ *such that* $f_0 = f | L$ *then there exists a homotopy* $F_t : K \to X$ *such that* $F_0 = f$ *and* $F_t | L = f_t | L$, $0 \leq t \leq 1$. (Reference: [SCHUBERT, p. 197]). □

As an application of (3.1) we get

(3.2) *If $L < K$ then the following assertions are equivalent:*

(1) $K \,\tilde{\searrow}\, L$
(2) *The inclusion map $i: L \subset K$ is a homotopy equivalence.*
(3) $\pi_n(K, L) = 0$ *for all* $n \leq \dim(K-L)$.

COMMENT ON PROOF: The implications (1) ⇒ (2) and (2) ⇒ (3) are elementary. The implication (3) ⇒ (1) is proved inductively, using (3) and the homotopy extension property to construct first a homotopy (rel L) of 1_K to a map $f_0: K \to K$ which takes K^0 into L, then to construct a homotopy (rel L) of f_0 to $f_1: K \to K$ such that $f_1(K^1) \subset L$, and so on. □

If K_0 and K_1 are CW complexes, a map $f: K_0 \to K_1$ is *cellular* if $f(K_0^n) \subset K_1^n$ for all n. More generally, if (K_0, L_0) and (K_1, L_1) are CW pairs, a map $f:(K_0,L_0) \to (K_1,L_1)$ is *cellular* if $f(K_0^n \cup L_0) \subset (K_1^n \cup L_1)$ for all n. Notice that this does not imply that $f|L_0: L_0 \to L_1$ is cellular. As a typical example, suppose that I^n is given a cell structure with exactly one n-cell and suppose that $f: I^n \to K$ is a characteristic map for some cell e. Then $f:(I^n, \partial I^n) \to (K, K^{n-1})$ is cellular while $f|\partial I^n$ need not be cellular.

If $f \simeq g$ and g is cellular then g is called a *cellular approximation to f.*

(3.3) (The cellular approximation theorem) *Any map between* CW *pairs, $f:(K_0,L_0) \to (K_1,L_1)$ is homotopic* (rel. L_0) *to a cellular map.* (Reference: [Schubert, p. 198]). □

If A is a closed subset of X and $f: A \to Y$ is a map then $X \cup_f Y$ is the identification space $[X \oplus Y \,/\, x = f(x) \text{ if } x \in A]$.

(3.4) *Suppose that $K_0 < K$ and $f: K_0 \to L$ is a map such that, given any cell e of $K - K_0$, $f(\bar{e} \cap K_0) \subset L^{n-1}$ where* $\dim e = n$. *Then $K \cup_f L$ is a* CW *complex whose cells are those of $K - K_0$ and those of L.* (*More precisely the cells of $K \cup_f L$ are of the form $q(e)$ where e is an arbitrary cell of $K - K_0$ or of L and $q: K \oplus L \to K \cup L$ is the identification map. We suppress q whenever possible*). □

Using (3.4) and the natural cell structure on $K \times I$ we get

(3.5) *If $f: K \to L$ is a cellular map then the mapping cylinder M_f is a* CW *complex with cells which are either cells of L or which are of the form $e \times 0$ or $e \times (0,1)$, where e is an arbitrary cell of K.* □

Combining (1.2), (3.2) and (3.5) we have

(3.6) *A cellular map $f: K \to L$ is a homotopy equivalence if and only if $M_f \,\tilde{\searrow}\, K$.* □

Cellular homology theory

If (K, L) is a CW pair, the *cellular chain complex* $C(K, L)$ is defined by letting $C_n(K, L) = H_n(K^n \cup L, K^{n-1} \cup L)$ and letting $d_n: C_n(K, L) \to C_{n-1}(K, L)$ be the boundary operator in the exact sequence for singular homology of the triple $(K^n \cup L, K^{n-1} \cup L, K^{n-2} \cup L)$.

$C_n(K, L)$ is usually thought of as "the free module generated by the n-cells of $K-L$". To make this precise, let us adopt, now and forever, standard orientations ω_n of $I^n (n = 0,1,2,\ldots)$ by choosing a generator ω_0 of $H_0(I^0)$ and stipulating that the sequence of isomorphisms

$$H_{n-1}(I^{n-1}, \partial I^{n-1}) \xrightarrow{\text{excision}} H_{n-1}(\partial I^n, J^{n-1}) \xleftarrow{\supset} H_{n-1}(\partial I^n) \xleftarrow{\partial} H_n(I^n, \partial I^n)$$

takes ω_{n-1} onto $-\omega_n$. (Here $I^{n-1} \equiv I^{n-1} \times 0$). If $\varphi_\alpha: I^n \to K$ is a characteristic map for $e_\alpha \in K-L$ we denote $\langle\varphi_\alpha\rangle = (\varphi_\alpha)_*(\omega_n)$ where $(\varphi_\alpha)_*: H_n(I^n, \partial I^n) \to H_n(K^n \cup L, K^{n-1} \cup L)$ is the induced map. Then the situation is described by the following two lemmas.

(3.7) *Suppose that a characteristic map φ_α is chosen for each n-cell e_α of $K-L$. Denote $K_j = K^j \cup L$. Then*

(a) $H_j(K_n, K_{n-1}) = 0$ *if* $j \neq n$
(b) $H_n(K_n, K_{n-1})$ *is free with basis* $\{\langle\varphi_\alpha\rangle | e^n_\alpha \in K-L\}$
(c) *If* c is a singular n-cycle of K mod L representing $\gamma \in H_n(K_n, K_{n-1})$ *and if* $|c|$ *does not include the* n-cell e_{α_0} *then* $n_{\alpha_0} = 0$ *in the expression* $\gamma = \sum_\alpha n_\alpha \langle\varphi_\alpha\rangle$. (Reference: [G. W. WHITEHEAD, p. 58] and [SCHUBERT, p. 300]). □

A cellular map $f:(K,L) \to (K',L')$ clearly induces a chain map $f_*: C(K,L) \to C(K',L')$ and thus a homomorphism, also called f_*, from $H(C(K,L))$ to $H(C(K',L'))$. Noting this, the cellular chain complex plays a role in the category of CW complexes analogous to that played by the simplicial chain complex in the simplicial category because of

(3.8) *There is a natural equivalence T between the "cellular homology" functor and the "singular homology" functor. In other words, for every* CW *pair (K,L) there is an isomorphism $T_{K,L}: H(C(K,L)) \to H(|K|, |L|)$, and* for every *cellular* map $f:(K,L) \to (K',L')$ *the following diagram commutes*

$$\begin{array}{ccc} H(C(K,L)) & \xrightarrow{T_{K,L}} & H(|K|, |L|) \\ \downarrow f_* & & \downarrow f_* \\ H(C(K',L')) & \xrightarrow{T_{K',L'}} & H(|K'|, |L'|) \end{array}$$

The isomorphism $T_{K,L}$ takes the homology class of a cycle $\sum_i n_i\langle\varphi_{\alpha_i}\rangle \in C_n(K, L)$ onto the homology class of the cycle $\sum_i n_i\bar\varphi_{\alpha_i} \in S_n(K, L)$, where $\bar\varphi_{\alpha_i}$ is a singular chain representing $\langle\varphi_{\alpha_i}\rangle$. (Reference: [G.W. WHITEHEAD, p. 65] and [SCHUBERT, p. 305]). □

(3.9) *Suppose that $f: K \to L$ is a cellular map with mapping cylinder M_f. Then $C(M_f,K)$ is naturally isomorphic to the chain complex $(\mathscr{C},\partial)$—"the mapping cone" of $f_*: C(K) \to C(L)$—which is given by*

$$\mathscr{C}_n = C_{n-1}(K) \oplus C_n(L)$$

$$\partial_n(x+y) = -d_{n-1}(x)+[f_*(x)+d_n'(y)], \quad x \in C_{n-1}(K), \quad y \in C_n(L)$$

where d and d' are the boundary operators in $C(K)$ and $C(L)$ respectively.

By "naturally isomorphic" we mean that, for each n, the isomorphism constructed algebraically realizes the correspondence between n-cells of $M_f - K$ and cells of $K^{n-1} \cup L^n$ given by $e^{n-1} \times (0,1) \leftrightarrow e^{n-1}$ and $u^n \leftrightarrow u^n$ (e^{n-1} a cell of K, u^n a cell of L).

PROOF OF (3.9): Let $\{e_\alpha\}$ be the cells of K and suppose that characteristic maps φ_α have been chosen. Then $(K \times I, K \times 0)$ is a CW pair with relative cells of the form $e_\alpha \times 1$ and $e_\alpha \times (0,1)$ possessing the obvious characteristic maps $\varphi_{\alpha,1}$ and $\varphi_\alpha \times 1_I$. If $\dim e_\alpha = n-1$, let $\langle\varphi_\alpha\rangle = \varphi_{\alpha,1*}(\omega_{n-1})$ and $\langle\varphi_\alpha\rangle \times I = (\varphi_\alpha \times 1_I)_*(\omega_n)$ be the corresponding basis elements of $C(K \times I, K \times 0)$. In general, if $c = \sum_i n_i\langle\varphi_{\alpha_i}\rangle$ is an arbitrary element of $C_{n-1}(K)$, set $c \times I = \sum_i n_i(\langle\varphi_{\alpha_i}\rangle \times I)$. In the product cell structure for I^n we have $\omega_n \in C_n(I^n)$ and (exercise—induction on n suggested) $d\omega_n = \sum_{j=1}^{n} (-1)^{n-j}$ $(i_{j,1*}\omega_{n-1} - i_{j,0*}\omega_{n-1}) \in C_{n-1}(I^n)$ where $i_{j,\varepsilon}: I^{n-1} \to I^n$ is the characteristic map $i_{j,\varepsilon}(t_1,\ldots,t_{n-1}) = (t_1,\ldots,t_{j-1},\varepsilon,t_j,\ldots,t_{n-1})$, $\varepsilon = 0,1$. This gives $d\omega_n = i_{n,1*}\omega_{n-1} - i_{n,0*}\omega_{n-1} - (d\omega_{n-1} \times I)$. Interpreted in $C(I^n, I^{n-1} \times 0)$ this becomes $d\omega_n = i_{n,1*}\omega_{n-1} - (d\omega_{n-1} \times I)$, and applying the chain map $(\varphi_\alpha \times 1_I)_*$ we get

$$d(\langle\varphi_\alpha\rangle \times I) = \langle\varphi_\alpha\rangle - (d\langle\varphi_\alpha\rangle \times I) \in C_{n-1}(K \times I, K \times 0).$$

Let $\{u_\beta\}$ be the cells of L, with characteristic maps ψ_β. Then $q_*: C(K \times I, K \times 0) \oplus C(L) \to C(M_f,K)$, and $C(M_f,K)$ has as basis—from the natural cell structure of M_f—the set

$$\{q_*(\langle\varphi_\alpha\rangle \times I) | e_\alpha \in K\} \cup \{q_*\langle\psi_\beta\rangle | u_\beta \in L\},$$

Define a degree-zero homomorphism $T: C(M_f,K) \to \mathscr{C}$ by stipulating that $T(q_*(\langle\varphi_\alpha\rangle \times I)) = \langle\varphi_\alpha\rangle$ and $T(q_*\langle\psi_\beta\rangle) = \langle\psi_\beta\rangle$. Notice that (with the obvious identifications) $Tq_*|C(K \times 1) = f_*: C(K) \to C(L)$ and $Tq_*(c \times I) = c$ for all $c \in C(K)$. Thus

$$\begin{aligned} Td[q_*(\langle\varphi_\alpha\rangle \times I)] &= Tq_*d[\langle\varphi_\alpha\rangle \times I] \\ &= Tq_*[\langle\varphi_\alpha\rangle - (d\langle\varphi_\alpha\rangle \times I)] \\ &= Tq_*(\langle\varphi_\alpha\rangle) - Tq_*(d\langle\varphi_\alpha\rangle \times I) \\ &= f_*\langle\varphi_\alpha\rangle - d\langle\varphi_\alpha\rangle \\ &= \partial\langle\varphi_\alpha\rangle = \partial T[q_*(\langle\varphi_\alpha\rangle \times I)] \end{aligned}$$

It follows trivially that T is a chain isomorphism. □

Covering spaces

We turn now to covering spaces. Connectivity of the base space will be assumed throughout this discussion

(3.10) *If K is a CW complex then K is locally contractible. Thus for any subgroup* $G \subset \pi_1(K)$ *there is a covering space* $p: E \to K$ *such that* $p_\#(\pi_1 E) = G$. *In particular K has a universal covering space.* (Reference: [SCHUBERT, p. 204]). □

We define $p: E \to K$ to be a *covering in the CW category* provided that p is a covering map and that E and K are CW complexes such that the image of every cell of E is a cell of K. By a *covering* we shall always mean a covering in the CW category if the domain is a CW complex. Nothing is lost in doing this because of

(3.11) *Suppose that K is a* CW *complex and* $p: E \to K$ *is a covering of K. Then*

$$\{\tilde{e}_\alpha | e_\alpha \in K, \tilde{e}_\alpha \text{ is a lift of } e_\alpha \text{ to } E\}$$

is a cell structure on E with respect to which E becomes a CW *complex. If* $\varphi_\alpha: I^n \to K$ *is a characteristic map for the cell* e_α, *if* $\tilde{e}_\alpha$ *is a lift of* e_α *and if* $\tilde{\varphi}_\alpha: I^n \to E$ *is a lift of* φ_α *such that* $\tilde{\varphi}_\alpha(x) \in \tilde{e}_\alpha$ *for some* $x \in \mathring{I}^n$, *then* $\tilde{\varphi}_\alpha$ *is a characteristic map for* $\tilde{e}_\alpha$. (Reference: [SCHUBERT, p. 251]). □

(3.12) *If* $p: E \to K$ *is a covering and* $f: K' \to K$ *is a cellular map which lifts to* $\tilde{f}: K' \to E$ *then* $\tilde{f}$ *is cellular. If f is a covering (in the CW category), so is* $\tilde{f}$. □

Since a covering which is also a homeomorphism is a cellular isomorphism, (3.12) implies that the universal covering space of K is unique up to cellular isomorphism.

(3.13) *Suppose that* (K, L) *is a pair of connected* CW *complexes and that* $p: \tilde{K} \to K$ *is the universal covering.* Let $\tilde{L} = p^{-1}L$. *If* $i_\#: \pi_1 L \xrightarrow{\subset} \pi_1 K$ *is an isomorphism then* $p|\tilde{L}: \tilde{L} \to L$ *is the universal covering of L. If, further,* $K \searrow L$ *then* $\tilde{K} \searrow \tilde{L}$.

PROOF: $\tilde{L}$ is a closed set which is the union of cells of $\tilde{K}$ (namely, the lifts of the cells of L). Thus $\tilde{L}$ is a subcomplex of $\tilde{K}$. Clearly $p|\tilde{L}$ is a covering of L. We shall show that, if $i_\#$ is an isomorphism, $\tilde{L}$ is connected and simply connected. Notice that, by the covering homotopy property, $p_\#: \pi_i(\tilde{K}, \tilde{L}) \cong \pi_i(K, L)$ for all $i \geq 1$. To see that $\tilde{L}$ is connected, notice that $\pi_1(K, L) = 0$ since we have exactness in the sequence

$$\pi_1(L) \xrightarrow{\cong} \pi_1(K) \to \pi_1(K, L) \to \pi_0(L) \xrightarrow{\cong} \pi_0(K).$$

Thus $\pi_1(\tilde{K}, \tilde{L}) = 0$. Hence by the connectedness of $\tilde{K}$ and the exactness of the sequence

$$0 = \pi_1(\tilde{K}, \tilde{L}) \to \pi_0(\tilde{L}) \to \pi_0(\tilde{K})$$

it follows that $\tilde{L}$ is connected.

$\tilde{L}$ is 1-connected because of the commutativity of the diagram

$$\begin{array}{ccc} \pi_1\tilde{L} & \longrightarrow & \pi_1\tilde{K} = 0 \\ \Big\downarrow {\scriptstyle 1-1} & & \Big\downarrow {\scriptstyle 0} \\ \pi_1 L & \xrightarrow{\cong} & \pi_1 K \end{array}$$

Hence $p: \tilde{L} \to L$ is the universal covering.

Finally, $K \rightsquigarrow L$ implies $\pi_i(K, L) = 0$ and hence $\pi_i(\tilde{K}, \tilde{L}) = 0$ for all $i \geq 1$. Thus $\tilde{K} \rightsquigarrow \tilde{L}$ by (3.2). □

(3.14) *Suppose that $f: K \to L$ is a cellular map between connected complexes such that $f_{\#}: \pi_1 K \to \pi_1 L$ is an isomorphism. If $\tilde{K}$, $\tilde{L}$ are universal covering spaces of K, L and $\tilde{f}: \tilde{K} \to \tilde{L}$ is a lift of f, then $M_{\tilde{f}}$ is a universal covering space of M_f.*

Exercise: Give a counter-example when $f_{\#}$ is not an isomorphism.

PROOF OF (3.14): $\tilde{f}$ is cellular and $M_{\tilde{f}} \searrow \tilde{L}$, so $M_{\tilde{f}}$ is a simply connected CW complex. Let $p: \tilde{K} \to K$ and $p': \tilde{L} \to L$ be the covering maps. Define $\alpha: M_{\tilde{f}} \to M_f$ by

$$\alpha[w, t] = [p(w), t], \qquad 0 \leq t \leq 1,\ w \in \tilde{K}$$

$$\alpha[z] = [p'(z)], \qquad z \in \tilde{L}$$

If $[w, 1] = [z]$ then $\tilde{f}(w) = z$, so $\alpha[w, 1] = [p(w), 1] = [fp(w)] = [p'\tilde{f}(w)] = [p'(z)]$. Hence α is well-defined. It is clearly continuous. Notice that $\alpha|(M_{\tilde{f}} - \tilde{L}) = \alpha|\tilde{K} \times [0, 1) = p \times 1_{[0,1)}$ and $\alpha|\tilde{L} = p'$. Thus $\alpha|M_{\tilde{f}} - \tilde{L})$ and $\alpha|\tilde{L}$ are covering maps, and α takes cells homeomorphically onto cells.

Let $\beta: \hat{M}_f \to M_f$ be the universal cover of M_f, with $\hat{K} = \beta^{-1}(K)$, $\hat{L} = \beta^{-1}(L)$. By (3.13), $\beta|\hat{L}: \hat{L} \to L$ is a universal covering. Since $f_{\#}: \pi_1 K \to \pi_1 L$ is an isomorphism, so, by (1.2), is $i_{\#}: \pi_1 K \to \pi_1 M_f$. Hence $\hat{K}$ is simply connected, using (3.13) again. But clearly $\beta|(\hat{M}_f - \hat{L}): \hat{M}_f - \hat{L} \to M_f - L$ is a covering and $\pi_i(M_f - L, K) = \pi_i(\hat{M}_f - \hat{L}, \hat{K}) = 0$ for all i. So $\hat{M}_f - \hat{L}$ is simply connected and $\beta|(\hat{M}_f - \hat{L})$ is a universal covering also.

Now let $\hat{\alpha}: M_{\tilde{f}} \to \hat{M}_f$ be a lift of α. By uniqueness of the universal covering spaces of $M_f - L$ and L, $\hat{\alpha}$ must take $M_{\tilde{f}} - \tilde{L}$ homeomorphically onto $\hat{M}_f - \hat{L}$ and $\tilde{L}$ homeomorphically onto $\hat{L}$. Thus $\hat{\alpha}$ is a continuous bijection. But it is clear that $\hat{\alpha}$ takes each cell e homeomorphically onto a cell $\hat{\alpha}(e)$. Then $\hat{\alpha}$ takes $\bar{e}$ bijectively, hence homeomorphically, onto $\hat{\alpha}(\bar{e})$. The latter is just $\overline{\hat{\alpha}(e)}$ because if φ is a characteristic map for e, $\hat{\alpha}\varphi$ is a characteristic map for $\hat{\alpha}(e)$, so that $\overline{\hat{\alpha}(e)} = \hat{\alpha}\varphi(I^n) = \hat{\alpha}(\bar{e})$. Since $M_{\tilde{f}}$ and $\hat{M}_f$ have the weak topology with respect to closed cells it follows that $\hat{\alpha}$ is a homeomorphism. Since $\beta\hat{\alpha} = \alpha$ it follows that α is a covering map. □

Consider now the cellular chain complex $C(\tilde{K}, \tilde{L})$, where $\tilde{K}$ is the universal covering space of K and $L < K$. Besides being a $\mathbb{Z}$-module with the properties given by (3.7) and (3.8), $C(\tilde{K}, \tilde{L})$ is actually a $\mathbb{Z}(G)$-module where G is the group of covering homeomorphisms of $\tilde{K}$ or, equivalently, the fundamental group of K. We wish to explain how this richer structure comes about.

Recall the definition: If G is a group and $\mathbb{Z}$ is the ring of integers then $\mathbb{Z}(G)$—*the integral group ring of G*—is the set of all finite formal sums $\sum_i n_i g_i$, $n_i \in \mathbb{Z}$, $g_i \in G$, with addition and multiplication given by

$$\sum_i n_i g_i + \sum_i m_i g_i = \sum_i (n_i + m_i) g_i$$

$$\left(\sum_i n_i g_i\right)\cdot\left(\sum_j m_j g_j\right) = \sum_{i,j} (n_i m_j)(g_i g_j)$$

One can similarly define $\mathbb{R}(G)$ for any ring $\mathbb{R}$.

Let $p: \tilde{K} \to K$ be the universal covering and let $G = \mathrm{Cov}(\tilde{K}) =$ [the set of all homeomorphisms $h: \tilde{K} \to \tilde{K}$ such that $ph = p$]. Suppose that $L < K$ and $\tilde{L} = p^{-1}L$. Each $g \in G$ is (3.12) a cellular isomorphism of $\tilde{K}$ inducing, for each n, the homomorphism $g_*: C_n(\tilde{K}, \tilde{L}) \to C_n(\tilde{K}, \tilde{L})$ and satisfying $dg_* = g_* d$ (where d is the boundary operator in $C(\tilde{K}, \tilde{L})$). Let us define an action of G on $C(\tilde{K}, \tilde{L})$ by $g \cdot c = g_*(c)$, $(g \in G,\ c \in C(\tilde{K}, \tilde{L}))$. Clearly $d(g \cdot c) = g \cdot (dc)$. Thus $C(\tilde{K}, \tilde{L})$ becomes a $\mathbb{Z}(G)$-complex if we define

$$\left(\sum_i n_i g_i\right)\cdot c = \sum_i n_i (g_i \cdot c) = \sum_i n_i (g_i)_*(c)$$

The following proposition shows that $C(\tilde{K}, \tilde{L})$ is a free $\mathbb{Z}(G)$-complex with a natural class of bases.

(3.15) *Suppose that $p: \tilde{K} \to K$ is the universal covering and that G is the group of covering homeomorphisms of $\tilde{K}$. Assume that $L < K$ and $\tilde{L} = p^{-1}L$. For each cell e_α of $K - L$, let a specific characteristic map $\varphi_\alpha: I^n \to K$ $(n = n(\alpha))$ and a specific lift $\tilde{\varphi}_\alpha: I^n \to \tilde{K}$ of φ_α be chosen. Then $\{\langle \tilde{\varphi}_\alpha \rangle | e_\alpha \in K - L\}$ is a basis for $C(\tilde{K}, \tilde{L})$ as a $\mathbb{Z}(G)$-complex.*

PROOF: Let $* = *_n$ be a fixed point of $\mathring{I}^n$ for each n. For each $y \in p^{-1}\varphi_\alpha(*)$, let $\hat{\varphi}_{\alpha,y}$ be the unique lift of φ_α with $\hat{\varphi}_{\alpha,y}(*) = y$. Since $p: \tilde{K} \to K$ is the universal covering, G acts freely and transitively on each fibre $p^{-1}(x)$. Thus each $\hat{\varphi}_{\alpha,y}$ is uniquely expressible as $\hat{\varphi}_{\alpha,y} = g \circ \tilde{\varphi}_\alpha$ for some $g \in G$ and $\{\hat{\varphi}_{\alpha,y} | y \in p^{-1}\varphi_\alpha(*)\} = \{g \circ \tilde{\varphi}_\alpha | g \in G\}$. But, by (3.7) and (3.11), $C(\tilde{K}, \tilde{L})$ is a free $\mathbb{Z}$-module with basis

$$\{\langle \hat{\varphi}_{\alpha,y} \rangle\} = \{\langle g \circ \tilde{\varphi}_\alpha \rangle\} = \{g_* \langle \tilde{\varphi}_\alpha \rangle\} = \{g \cdot \langle \tilde{\varphi}_\alpha \rangle\}$$

where g varies over G and φ_α varies over the given characteristic maps for $K - L$. Thus each $c \in C(\tilde{K}, \tilde{L})$ is uniquely representable as a finite sum

$$\begin{aligned} c &= \sum_{i,\alpha} n_{i,\alpha}(g_i \cdot \langle \tilde{\varphi}_\alpha \rangle) \\ &= \sum_\alpha \left(\sum_i n_{i,\alpha} g_i\right) \cdot \langle \tilde{\varphi}_\alpha \rangle \\ &= \sum_\alpha r_\alpha \langle \tilde{\varphi}_\alpha \rangle,\ r_\alpha \in \mathbb{Z}(G) \end{aligned}$$

Therefore $\{\langle \varphi_\alpha \rangle | e_\alpha$ is a cell of $K - L\}$ is a basis of $C(\tilde{K}, \tilde{L})$ as a $\mathbb{Z}(G)$-module. □

The fundamental group and the group of covering transformations

If we choose base points $x \in K$ and $\tilde{x} \in p^{-1}(x)$ then there is a standard identification of the group of covering transformations G with $\pi_1 K = \pi_1(K,x)$. Because of its importance in the sequel, we review this in some detail.

For each $\alpha:(I,\dot{I}) \to (K,x)$, let $\tilde{\alpha}$ be the lift of α with $\tilde{\alpha}(0) = \tilde{x}$. Let $g_{[\alpha]}:\tilde{K} \to \tilde{K}$ be the unique covering homeomorphism such that $g_{[\alpha]}(\tilde{x}) = \tilde{\alpha}(1)$. We claim that, if $y \in \tilde{K}$ and if $\omega:(I,0,1) \to (\tilde{K},\tilde{x},y)$ is any path, then

$$g_{[\alpha]}(y) = \widetilde{\alpha * p\omega}\,(1)$$

where $p\omega$ is the composition of ω and p, and "$*$" represents concatenation of loops. To see this, note that $\widetilde{\alpha * p\omega}(1) = \widehat{p\omega}(1)$ where $\widehat{p\omega}$ is the unique lift of $p\omega$ with $\widehat{p\omega}(0) = \tilde{\alpha}(1)$. But $g_{[\alpha]}(\widetilde{p\omega}(0)) = g_{[\alpha]}(\tilde{x}) = \tilde{\alpha}(1)$, so $g_{[\alpha]} \circ \widetilde{p\omega}$ is such a lift. Hence $\widehat{p\omega} = g_{[\alpha]} \circ \widetilde{p\omega}$ and

$$\widetilde{\alpha * p\omega}(1) = g_{[\alpha]}(\widetilde{p\omega}(1)) = g_{[\alpha]}(y).$$

The function $\theta = \theta(x,\tilde{x}):\pi_1 K \to G$, given by $[\alpha] \to g_{[\alpha]}$, is an isomorphism. For example, it is a homomorphism because, for arbitrary $[\alpha]$, $[\beta] \in \pi_1 K$, we have (by the preceding paragraph)

$$\begin{aligned} g_{[\alpha]}(g_{[\beta]}(\tilde{x})) &= g_{[\alpha]}(\tilde{\beta}(1)) \\ &= \widetilde{\alpha * p\tilde{\beta}}(1) \\ &= \widetilde{\alpha * \beta}(1) \\ &= g_{[\alpha][\beta]}(\tilde{x}) \end{aligned}$$

Hence $g_{[\alpha]} \circ g_{[\beta]} = g_{[\alpha][\beta]}$, since they agree at a point.

Suppose that $p:\tilde{K} \to K$ and $p':\tilde{L} \to L$ are universal coverings with $p(\tilde{x}) = x$ and $p'(\tilde{y}) = y$, and that G_K and G_L are the groups of covering transformations. Then any map $f:(K,x) \to (L,y)$ induces a unique map $f_\#:G_K \to G_L$ such that the diagram

$$\begin{array}{ccc} G_K & \xrightarrow{\ f_\#\ } & G_L \\ \uparrow{\scriptstyle \theta(x,\tilde{x})} & & \uparrow{\scriptstyle \theta(y,\tilde{y})} \\ \pi_1(K,x) & \xrightarrow{\ f_\#\ } & \pi_1(L,y) \end{array}$$

commutes. (We believe that it aids the understanding to call both maps $f_\#$.) This map satisfies

(3.16) *If* $g \in G_K$ *and* $\tilde{f}:(\tilde{K},\tilde{x}) \to (\tilde{L},\tilde{y})$ *covers* f, *then* $f_\#(g) \circ \tilde{f} = \tilde{f} \circ g$.

PROOF: Since these maps both cover f, it suffices to show that they agree at a single point—say $\tilde{x}$. So we must show that $(f_\#(g))(\tilde{y}) = \tilde{f}g(\tilde{x})$. Letting α be a loop such that $[\alpha]$ corresponds to g under $\theta(x,\tilde{x})$, we have

$$\begin{aligned}
\tilde{f}g(\tilde{x}) &= \tilde{f}\tilde{\alpha}(1) \\
&= (\widetilde{f\circ\alpha})(1), \quad \text{since} \quad \tilde{f}\tilde{\alpha}(0) = \tilde{y} = \widetilde{f\alpha}(0) \\
&= (\theta(y,\tilde{y})(f_\#[\alpha]))(\tilde{y}), \quad \text{where} \quad f_\#:\pi_1(K,x) \to \pi_1(L,y) \\
&= ((\theta(y,\tilde{y})f_\#\theta(x,\tilde{x})^{-1})(g))(\tilde{y}) \\
&= (f_\#(g))(\tilde{y}). \quad \square
\end{aligned}$$

Chapter II

A Geometric Approach to Homotopy Theory

From here on all CW complexes mentioned will be assumed finite unless they occur as the covering spaces of given finite complexes.

§4. Formal deformations

Suppose that (K, L) is a finite CW pair. Then $K \overset{e}{\searrow} L$—i.e., *K collapses to L by an elementary collapse*—iff

(1) $K = L \cup e^{n-1} \cup e^n$ where e^n and e^{n-1} are not in L,

(2) there exists a ball pair $(Q^n, Q^{n-1}) \approx (I^n, I^{n-1})$ and a map $\varphi: Q^n \to K$ such that

(a) φ is a characteristic map for e^n

(b) $\varphi|Q^{n-1}$ is a characteristic map for e^{n-1}

(c) $\varphi(P^{n-1}) \subset L^{n-1}$, where $P^{n-1} \equiv \mathrm{Cl}(\partial Q^n - Q^{n-1})$.

In these circumstances we also write $L \overset{e}{\nearrow} K$ and say that *L expands to K by an elementary expansion*. It will be useful to notice that, if (2) is satisfied for one ball pair (Q^n, Q^{n-1}), it is satisfied for any other such ball pair, since we need only compose φ with an appropriate homeomorphism.

Geometrically, the elementary expansions of L correspond precisely to the attachings of a ball to L along a face of the ball by a map which is almost, but not quite, totally unrestricted. For, if we set $\varphi_0 = \varphi|P^{n-1}$ in the above definition, then $\varphi_0:(P^{n-1}, \partial P^{n-1}) \to (L^{n-1}, L^{n-2})$ and

$$(K, L) \cong (L \underset{\varphi_0}{\cup} Q^n, L).$$

Conversely, given L, any map $\varphi_0:(P^{n-1}, \partial P^{n-1}) \to (L^{n-1}, L^{n-2})$ determines an elementary expansion. To see this, set $K = L \underset{\varphi_0}{\cup} Q^n$. Let $\varphi: L \oplus Q^n \to K$ be the quotient map and define $\varphi(\mathring{Q}^{n-1}) = e^{n-1}$, $\varphi(\mathring{Q}^n) = e^n$. Then $K = L \cup e^{n-1} \cup e^n$ is a CW complex and $L \overset{e}{\nearrow} K$.

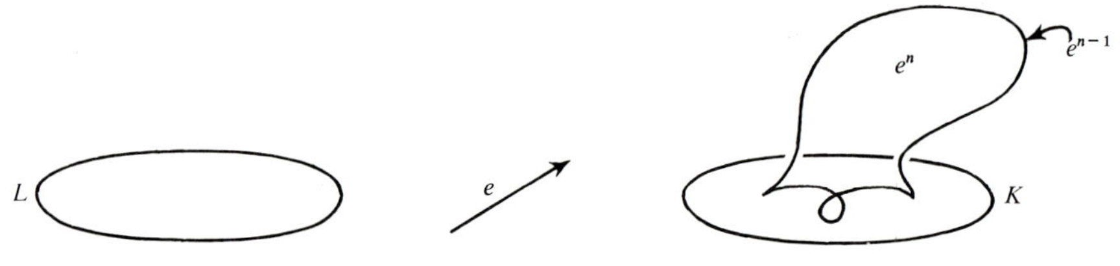

(4.1) *If* $K \searrow^e L$ *then* (a) *there is a cellular strong deformation retraction* $D: K \to L$ *and* (b) *any two strong deformation retractions of K to L are homotopic rel L.*

PROOF. Let $K = L \cup e^{n-1} \cup e^n$. By hypothesis there is a map $\varphi_0: I^{n-1} \to L^{n-1}$ such that $(K, L) \approx (L \underset{\varphi_0}{\cup} I^n, L)$. But $L \underset{\varphi_0}{\cup} I^n$ is just the mapping cylinder of φ_0. Hence, by (1.1) and its proof there is a strong deformation retraction $D: K \to L$ such that $D(\overline{e^n}) = \varphi_0(I^{n-1}) \subset L^{n-1}$. Clearly D is cellular.

If D_1 and $D_2: K \to L$ are two strong deformation retractions and $i: L \subset K$ then $iD_1 \simeq 1_K \simeq iD_2$ rel L. So $D_1 = D_1 i D_1 \simeq D_1 i D_2 = D_2$. □

We write $K \searrow L$ (*K collapses to L*) and $L \nearrow K$ (*L expands to K*) iff there is a finite sequence (possibly empty) of elementary collapses

$$K = K_0 \searrow^e K_1 \searrow^e \dots \searrow^e K_q = L.$$

A finite sequence of operations, each of which is either an elementary expansion or an elementary collapse is called a *formal deformation.* If there is a formal deformation from K to L we write $K \curlywedge L$. Clearly then, $L \curlywedge K$. K and L are then said to *have the same simple-homotopy type.* If K and L have a common subcomplex K_0, no cell of which is ever removed during the formal deformation, we write $K \curlywedge L$ rel K_0.

Suppose that $K = K_0 \to K_1 \to \dots \to K_q = L$ is a formal deformation. Define $f_i: K_i \to K_{i+1}$ by letting f_i be the inclusion map if $K_i \nearrow^e K_{i+1}$ and, (4.1), letting f_i be any cellular strong deformation retraction of K_i onto K_{i+1} if $K_i \searrow^e K_{i+1}$. Then $f = f_{q-1} \dots f_1 f_0$ is called a *deformation.* It is a cellular homotopy equivalence which is uniquely determined, up to homotopy, by the given formal deformation. If $K' < K$ and $f = f_{q-1} \dots f_0: K \to L$ is a deformation with each $f_i | K' = 1$ (so $K \curlywedge L$ rel K'), then we say that f is a *deformation rel K'.*

Finally, we define a *simple-homotopy equivalence* $f: K \to L$ to be a map which is homotopic to a deformation. f is a *simple-homotopy equivalence rel K'* if it is homotopic, rel K', to a deformation rel K'.

Some natural conjectures are

(I) If $f: K \to L$ is a homotopy equivalence then f is a simple-homotopy equivalence.

(II) If there exists a homotopy equivalence from K to L then there exists a simple-homotopy equivalence.

In general, both conjectures are false.[3] But in many special cases (e.g., if $\pi_1 L = 0$ or $\mathbb{Z}$ (integers)) both conjectures are true. And for some complexes L, **(I)** is false while **(II)** is true.

In the pages ahead, we shall concentrate on **(I)**—or, rather, on the equivalent conjecture **(I')** which is introduced in §5. Roughly, we will follow WHITEHEAD's path. We try to prove that **(I)** is true, run into an obstruction,

---

[3] See (24.1) and (24.4).

get some partial results, start all over and algebraicize the theory, and finally end up with a highly sophisticated theory which is, in the light of its evolution, totally natural.

Exercises:

4.A. If $K \searrow L$ then any given sequence of elementary collapses can be reordered to yield a sequence $K = K_0 \searrow^{e} K_1 \searrow^{e} \dots \searrow^{e} K_q = L$ with $K_i = K_{i+1} \cup e^{n_i} \cup e^{n_i - 1}$ where $n_0 \geq n_1 \geq \dots \geq n_{q-1}$.

4.B. If K is a contractible 1-dimensional finite CW-complex and x is any 0-cell then $K \searrow x$.

4.C. If $K \searrow x$ for some $x \in K^0$ then $K \searrow y$ for all $y \in K^0$.

4.D. If $K \curlywedge L$ then there are CW complexes P and L' such that $K \nearrow P \searrow L' \cong L$. (In essence: all the expansions can be done first.)

§5. Mapping cylinders and deformations

In this section we introduce some of the important facts relating mapping cylinders and formal deformations. The section ends by applying these facts to get a reformulation of conjecture **I** of §4.

(5.1) *If $f: K \to L$ is a cellular map and if $K_0 < K$ then $M_f \searrow M_{f|K_0}$.*

PROOF: Let $K = K_0 \cup e_1 \cup \dots \cup e_r$ where the e_i are the cells of $K - K_0$ arranged in order of increasing dimension. Then $K_i = K_0 \cup e_1 \cup \dots \cup e_i$ is a subcomplex of K. We set $M_i = M_{f|K_i}$ and claim that $M_i \searrow^{e} M_{i-1}$ for all i. For let φ_i be a characteristic map for e_i and let $q: (K_i \times I) \oplus L \to M_i$ be the quotient map. Then $M_i = M_{i-1} \cup e_i \cup (e_i \times (0, 1))$ and $q \circ (\varphi_i \times 1): I^{n_i} \times I \to M_i$ is a characteristic map for $(e_i \times (0, 1))$ which restricts on $I^{n_i} \times 0$ to a characteristic map for e_i. Clearly the complement of $I^{n_i} \times 0$ in $\partial(I^{n_i} \times I)$ gets mapped into $M_{i-1}^{n_i}$. Hence $M_i \searrow^{e} M_{i-1}$. Therefore $M_f \searrow M_{f|K_0}$. □

Corollary (5.1A): *If $f: K \to L$ is cellular then $M_f \searrow L$.* □

Corollary (5.1B): *If $K_0 < K$ then $(K \times I) \searrow (K_0 \times I) \cup (K \times i)$, $i = 0$* or 1. □

Corollary (5.1C): *If $K_0 < K$ and vK is the cone on K then $vK \searrow vK_0$.* □

Since we shall often pass from given CW complexes to isomorphic complexes without comment, we give the following lemma at the very outset.

(5.2): (a) *If (K, K_1, K_2) is a triple which is CW isomorphic to (J, J_1, J_2) and if $K \curlywedge K_1$ rel K_2 then $J \curlywedge J_1$ rel J_2.*

(b) *If K_1, K_2 and L are CW complexes with $L < K_1$* and *$L < K_2$ and if $h: K_1 \to K_2$ is a CW isomorphism such that $h|L = 1$ then $K_1 \curlywedge K_2$ rel L.*

PROOF: (a) is trivial and we omit the proof. To prove (b) it suffices to consider the special case where $(K_1 - L) \cap (K_2 - L) = \varnothing$. For if this is not the case we can (by renaming some points) construct a pair (K, L) and isomorphisms $h_i : K \to K_i$, $i = 1, 2$, such that $(K - L) \cap (K_i - L) = \varnothing$ and such that $h_i | L = 1$. Then, by the special case, $K_1 \mathrel{\nearrow\!\!\!\!\searrow} K \mathrel{\nearrow\!\!\!\!\searrow} K_2$, rel L.

Consider the mapping cylinder M_h. By (5.1),

$$M_h \searrow M_{h|L} = (L \times I) \cup (K_2 \times 1),$$

and, h being a CW isomorphism, the same proof can be used to collapse from the other end and get $M_h \searrow (L \times I) \cup (K_1 \times 0)$. Now let $\bar{M}_h$ be gotten from M_h by identifying $(x, t) = x$ if $x \in L$, $0 \le t \le 1$. Since $(K_1 - L) \cap (K_2 - L) = \varnothing$, we may (by taking an appropriate copy of $\bar{M}_h$) assume that K_1 and K_2 themselves, and not merely copies of them are contained in the two ends of $\bar{M}_h$. Then the collapses of M_h (rel $L \times I$) may be performed in this new context, since[4] $M_h - (L \times I)$ is isomorphic to $\bar{M}_h - L$, to yield $K_1 \nearrow \bar{M}_h \searrow K_2$ rel L. □

If we let $f : L \times I \to L$ be the natural projection, the argument in the last sentence is a special case of:

(5.3) (The relativity principle.) *Suppose that $L_1 < K$ and $f : L_1 \to L_2$ is a cellular map. If $K \mathrel{\nearrow\!\!\!\!\searrow} J$ rel L_1, then $K \cup_f L_2 \mathrel{\nearrow\!\!\!\!\searrow} J \cup_f L_2$ rel L_2 (by the "same" sequence of expansions and collapses).*

REMARK: In forming $K \cup_f L_2$ and $J \cup_f L_2$ one uses a "copy" of L_2 disjoint from K and J. By (5.2a) it doesn't matter which copy. In particular if f is an inclusion map we have as corollary:

(5.3′): *Suppose that $K \cup L_2$ and $J \cup L_2$ are CW complexes, with subcomplexes K, L_2 and J, L_2 respectively, and suppose that $K \cap L_2 = J \cap L_2 = L_1$. If $K \mathrel{\nearrow\!\!\!\!\searrow} J$ rel L_1 then $K \cup L_2 \mathrel{\nearrow\!\!\!\!\searrow} J \cup L_2$ rel L_2.*

PROOF of (5.3): Suppose that $K = K_0 \to K_1 \to \ldots \to K_p = J$ is a sequence of elementary deformations rel L_1. Let $q_i : K_i \oplus L_2 \to K_i \cup_f L_2$ be the quotient maps $(0 \le i \le p)$. If $K_{i \pm 1} \nearrow K_i = K_{i \pm 1} \cup e^{n-1} \cup e^n$, and $\varphi : I^n \to K_i$ is a characteristic map for e^n restricting to a characteristic map $\varphi | I^{n-1}$ for e^{n-1} then $q_i \varphi$ and $q_i(\varphi | I^{n-1})$ are characteristic maps for $q_i(e^n)$ and $q_i(e^{n-1})$, since $q_i | K_i - L_1$ is a homeomorphism and f is cellular. Thus

$$(K_{i \pm 1} \cup_f L_2) \nearrow (K_i \cup_f L_2) = (K_{i \pm 1} \cup_f L_2) \cup q_i(e^{n-1}) \cup q_i(e^n).$$

The result follows by induction on the number of elementary deformations. □

(5.4) *If $f : K \to L$ is a cellular map and $K \searrow K_0$ then $M_f \searrow K \cup M_{f|K_0}$.*

PROOF: Suppose that $K = K_p \searrow^{e} K_{p-1} \searrow \ldots \searrow^{e} K_0$. For fixed i let $K_{i+1} = K_i \cup (e^{n-1} \cup e^n)$ and let $\varphi : (I^n, I^{n-1}) \to (\overline{e^n, e^{n-1}})$ be an appropriate

[4] This is spelled out in the next proof.

characteristic map. Then

$$K \cup M_{f|K_{i+1}} = K \cup M_{f|K_i} \cup [e^{n-1} \times (0, 1) \cup e^n \times (0, 1)].$$

Then, q being the quotient map, $q \circ (\varphi \times 1):(I^n \times I, I^{n-1} \times I) \to K \cup M_{f|K_{i+1}}$ gives characteristic maps for these cells and meets the specifications for an elementary collapse. Hence $K \cup M_{f|K_{i+1}} \searrow^{e} K \cup M_{f|K_i}$. The result follows by induction. □

(5.5) *If f, $g:K \to L$ are homotopic cellular maps then $M_f \mathrel{\wedge\!\!\!\searrow} M_g$ rel $K \cup L$.*

PROOF: Let $F:K \times I \to L$ be a homotopy with $F_0 = f$ and $F_1 = g$. By the cellular approximation theorem we may assume that F is cellular. Then, by (5.4),

$$M_{F_0} \cup (K \times I) \nearrow M_F \searrow M_{F_1} \cup (K \times I)$$

since $(K \times I) \searrow K \times i\,(i = 0, 1)$. Now let $\pi:K \times I \to K$ be the natural projection and let $M = M_F \underset{\pi}{\cup} K$. By the relativity principle (5.3) the above deformation gives

$$M_f \nearrow M \searrow M_g \; rel\; K \cup L. \quad \square$$

(5.6) *If $f:K_1 \to K_2$ and $g:K_2 \to K_3$ are cellular maps then $M_{gf} \mathrel{\wedge\!\!\!\searrow} M_f \cup M_g$ rel $(K_1 \cup K_3)$ where $M_f \cup M_g$ is the disjoint union of M_f and M_g sewn together by the identity map on K_2.*

PROOF: Let $F = gp:M_f \to K_3$ where $p:M_f \to K_2$ is the natural retraction. Then F is a cellular map, $F|K_1 = gf$, and $F|K_2 = g$. Since $M_f \searrow K_2$ by (5.1A), it follows from (5.4) that $M_F \searrow M_f \cup M_g$. On the other hand, since $K_1 < M_f$, (5.1) implies that $M_F \searrow M_{gf}$. Thus $M_{gf} \nearrow M_F \searrow M_f \cup M_g$, where all complexes involved contain $K_1 \cup K_3$. □

More generally we have

(5.7) *If $K_1 \xrightarrow{f_1} K_2 \xrightarrow{f_2} \dots \xrightarrow{f_{q-1}} K_q$ is a sequence of cellular maps and $f = f_{q-1} \dots f_1$ then $M_f \mathrel{\wedge\!\!\!\searrow} M_{f_1} \cup M_{f_2} \cup \dots \cup M_{f_{q-1}}$, rel $(K_1 \cup K_q)$, where this union is the disjoint union of the M_{f_i} with the range of one trivially identified to the domain of the next.*

PROOF: This is trivial if $q = 2$. Proceeding inductively, set $g = f_{q-1} \dots f_3 f_2$ and assume $M_g \mathrel{\wedge\!\!\!\searrow} M_{f_2} \cup \dots \cup M_{f_{q-1}}$ *rel* $(K_2 \cup K_q)$. Then by (5.6) and (5.3′)

$$\begin{aligned} M_f = M_{gf_1} &\mathrel{\wedge\!\!\!\searrow} M_{f_1} \cup M_g, \; rel\; K_1 \cup K_q \\ &\mathrel{\wedge\!\!\!\searrow} M_{f_1} \cup (M_{f_2} \cup \dots \cup M_{f_{q-1}}), \; rel\; M_{f_1} \cup K_q. \quad \square \end{aligned}$$

(5.8) *Given a mapping $f:K \to L$, the following are equivalent statements:*

(a) *f is a simple-homotopy equivalence.*

(b) *There exists a cellular approximation g to f such that $M_g \mathrel{\wedge\!\!\!\searrow} K$, rel K.*

(c) *For any cellular approximation g to f, $M_g \mathrel{\wedge\!\!\!\searrow} K$, rel K.*

PROOF: (a) $\Rightarrow$ (b): By the definition of a simple-homotopy equivalence, there is a formal deformation

$$K = K_0 \to K_1 \to \dots \to K_q = L$$

such that f is homotopic to any deformation associated with this formal deformation. Let $g = g_{q-1} \dots g_1 g_0$ be such a deformation, where $g_i: K_i \to K_{i+1}$. Notice that, for all i, $M_{g_i} \searrow \operatorname{dom} g_i = K_i$.
For if $K_i \overset{e}{\nearrow} K_{i+1}$, then

$$M_{g_i} = (K_i \times I) \underset{g_i}{\cup} K_{i+1} \searrow (K_i \times I) \searrow (K_i \times 0) \equiv K_i$$

and if $K_i \overset{e}{\searrow} K_{i+1}$ then, by (5.4)

$$M_{g_i} \searrow M_{g_i|K_{i+1}} \cup K_i = (K_{i+1} \times I) \cup (K_i \times 0) \searrow K_i \times 0 \equiv K_i.$$

Thus

$$M_g \mathrel{\wedge\!\!\!\searrow} M_{g_0} \cup \dots \cup M_{g_{q-1}} \text{ rel } K_0, \qquad \text{by (5.7)}$$

$$\searrow (M_{g_0} \cup \dots \cup M_{g_{q-2}}) \searrow \dots \searrow M_{g_0} \searrow K_0 = K.$$

(b) $\Rightarrow$ (c): Suppose that g is a cellular approximation to f such that $M_g \mathrel{\wedge\!\!\!\searrow} K \text{ rel } K$ and that g' is any cellular approximation to f. Then, by (5.5), $M_{g'} \mathrel{\wedge\!\!\!\searrow} M_g \mathrel{\wedge\!\!\!\searrow} K \text{ rel } K$.
(c) $\Rightarrow$ (a): Let g be any cellular approximation to f. By hypothesis $M_g \mathrel{\wedge\!\!\!\searrow} K$, rel K. Thus the inclusion map $i: K \subset M_g$ is a deformation. Also the collapse $M_g \searrow L$ determines a deformation $P: M_g \to L$. Since any two strong deformation retractions are homotopic, P is homotopic to the natural projection $p: M_g \to L$. So $f \simeq g = pi \simeq Pi =$ deformation. Therefore f is a simple-homotopy equivalence. $\square$

(5.9) (The simple-homotopy extension theorem). *Suppose that $X < K_0 < K$ is a CW triple and that $f: K_0 \to L_0$ is a cellular simple-homotopy equivalence such that $f|X = 1$. Let $L = K \underset{f}{\cup} L_0$. Then there is a simple-homotopy equivalence $F: K \to L$ such that $F|K_0 = f$. Also $K \mathrel{\wedge\!\!\!\searrow} L$ rel X.*

PROOF: Let $F: K \to L$ be the restriction to K of the quotient map $K \oplus L_0 \to L$. Then $M_F = (K \times I) \underset{q}{\cup} M_f$ where $q: K_0 \times I \to M_f$ is also the restriction of a quotient map. But $K \times I \searrow (K_0 \times I) \cup (K \times 0)$, so

$$M_F \searrow M_f \cup (K \times 0) \equiv M_f \cup K, \qquad \text{by (5.3)}$$

$$\mathrel{\wedge\!\!\!\searrow} K \text{ rel } K, \qquad \text{by (5.8) and (5.3}').$$

Clearly $F|K_0 = f$ and, by (5.8) again, F is a simple-homotopy equivalence.
The last assertion of the theorem is true because

$$K \mathrel{\wedge\!\!\!\searrow} M_F \searrow M_{F|X} = (X \times I \cup L) \nearrow L \times I \searrow L \times 0 \equiv L$$

and this is all done rel $X = X \times 0$.[5] $\square$

---

5 The reader who is squeamish about "$L \times 0 \equiv L$" may invoke (5.2b).

In the light of (5.8), Conjecture **(I)** of §4 is equivalent to

(I′): *If* (X, Y) *is a CW pair and* $X \simeq Y$ *then* $X \nearrow\!\!\searrow Y$ *rel* Y.

For, assuming **(I′)**, suppose that $f: K \to L$ is a cellular homotopy equivalence. By (1.2), $M_f \simeq K$. Hence by **(I′)**, $M_f \nearrow\!\!\searrow K$ rel K; and by (5.8) f is a simple-homotopy equivalence, proving **(I)**. Conversely, assuming **(I)**, suppose that $X \simeq Y$—i.e., $i: Y \subset X$ is a homotopy equivalence. Then by **(I)**, i is a (cellular) simple-homotopy equivalence, so (5.8) implies that $M_i \nearrow\!\!\searrow Y$ rel Y. Therefore

$$X = X \times 0 \nearrow X \times I = M_{1_X} \searrow M_i \nearrow\!\!\searrow Y \text{ rel } Y,$$

proving **(I′)**.

We turn our attention therefore to Conjecture **(I′)** and (changing notation) to CW pairs (K, L) such that $K \simeq L$.

§6. The Whitehead group of a CW complex[6]

For a given finite CW complex, L, we wish to put some structure on the class of CW pairs (K, L) such that $K \simeq L$. We do so in this section, thus giving the first hint that our primitive geometry can be richly algebraicized.

If (K, L) and (K', L) are homotopically trivial CW pairs, define $(K, L) \sim (K', L)$ iff $K \nearrow\!\!\searrow K'$ rel L. This is clearly an equivalence relation and we let $[K, L]$ denote the equivalence class of (K, L). An addition of equivalence classes is defined by setting

$$[K, L] + [K', L] = [K \cup_L K', L]$$

where $K \cup_L K'$ is the disjoint union of K and K' identified by the identity map on L. {By 5.2 it doesn't matter which "disjoint union of K and K' identified . . ." we take. Also by (5.2) the equivalence classes form a set, since the isomorphism classes of finite CW complexes can easily be seen to have cardinality $\leq 2^c$.} The *Whitehead group of* L is defined to be the set of equivalence classes with the given addition and is denoted $Wh(L)$.

(6.1) $Wh(L)$ *is a well-defined abelian group.*

PROOF: A strong deformation retraction of K to L and one of K' to L combine trivially to give one of $K \cup_L K'$ to L. Thus $[K \cup_L K', L]$ is an element of $Wh(L)$ if $[K, L]$ and $[K', L]$ are. Moreover, if $[K, L] = [J, L]$, then $K \cup_L K' \nearrow\!\!\searrow J \cup_L K'$ rel L by (5.3′), so $[K \cup_L K', L] = [J \cup_L K', L]$. Similarly, if $[K', L] = [J', L]$, then $[J \cup_L K', L] = [J \cup_L J', L]$. Thus the addition is well defined.

[6] The viewpoint of this section has recently been arrived at by many people independently. It is interesting to compare [Stöcker], [Siebenmann], [Farrell-Wagoner], [Eckmann-Maumary] and the discussion here.

That the addition is associative and commutative follows from the fact that the union of sets has these properties.

The element $[L, L]$ is an identity, denoted by 0.

If $[K, L] \in Wh(L)$, let $D: K \to L$ be a cellular strong deformation retraction. Let $2M_D$ consist of two copies of the mapping cylinder M_D, identified by the identity on K. Precisely, let $2M_D = K \times [-1, 1]$ with the identifications $(x, -1) = (D(x), -1)$ and $(x, 1) = D(x)$ for all $x \in K$. We claim that $[2M_D, L] = -[K, L]$.

A picture of $2M_D \underset{L}{\cup} K$:

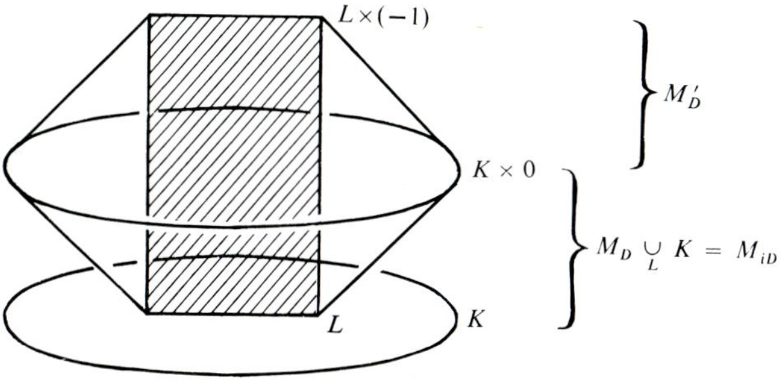

Proof of the claim:

$$
\begin{aligned}
[2M_D, L] + [K, L] &= [2M_D \underset{L}{\cup} K, L] \\
&= [(M_D \underset{L}{\cup} K) \cup M'_D, L] \\
&= [M_{iD} \cup M'_D, L] \quad \text{where} \quad i: L \subsetneq K.
\end{aligned}
$$

[But $iD \simeq 1_K$, so by (5.5), $M_{iD} \nearrow\!\!\!\searrow K \times I$ rel $(K \times 0) \cup K$. So by (5.3′) we have]

$$
\begin{aligned}
&= [K \times I \cup M'_D, L] \\
&= [L \times I \cup M'_D, L] && \text{since} \quad K \times I \searrow (L \times I \cup K \times 0) \\
&= [L \times [-1, 1], L] && \text{since} \quad M'_D \searrow L \times [-1, 0] \\
&= [L, L] = 0 && \text{since} \quad L \times [-1, 1] \searrow L \equiv L \times 1.
\end{aligned}
$$

In pictures, these equations represent

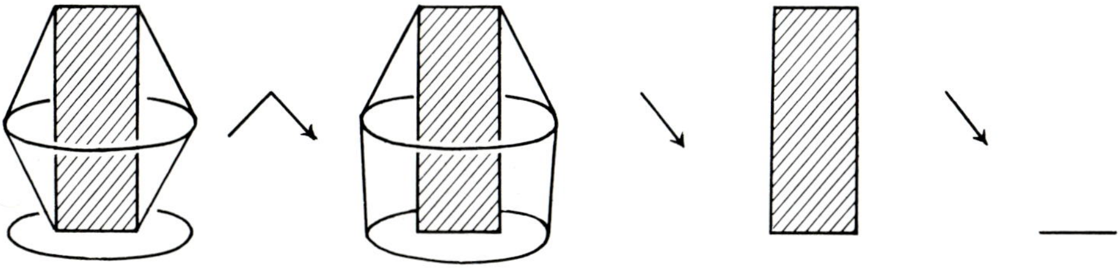

This completes the proof. □

If $f : L_1 \to L_2$ is a cellular map, we define $f_* : Wh(L_1) \to Wh(L_2)$ by

$$f_* [K, L_1] = [K \underset{L_1}{\cup} M_f, L_2]$$

or

$$f_* [K, L_1] = [K \underset{f}{\cup} L_2, L_2]$$

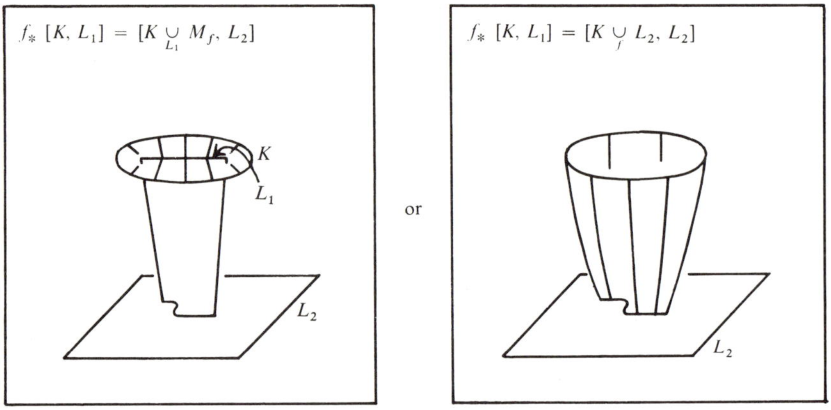

These definitions are equivalent because the natural projection $p : M_f \to L_2$ is a simple-homotopy equivalence with $p|L_2 = 1$ which, by (5.9), determines the deformation

$$(M_f \underset{L_1}{\cup} K) \searrow (M_f \underset{L_1}{\cup} K) \underset{p}{\cup} L_2 = K \underset{f}{\cup} L_2, \text{rel } L_2.$$

It follows directly from the second definition that f_* is a group homomorphism. From the first definition and (5.6) it follows directly that $g_* f_* = (gf)_*$. Leaving these verifications to the reader we now have

(6.2) *There is a covariant functor from the category of finite* CW *complexes and cellular maps to the category of abelian groups and group homomorphisms given by* $L \mapsto Wh(L)$ *and* $(f : L_1 \to L_2) \mapsto (f_* : Wh(L_1) \to Wh(L_2))$. *Moreover if* $f \simeq g$ *then* $f_* = g_*$.

PROOF: The reader having done his duty, we need only verify that if $f \simeq g$ then $f_* = g_*$. But this is immediate from the first definition of induced map and (5.5). $\square$

We can now define the *torsion* $\tau(f)$ of a cellular homotopy equivalence $f : L_1 \to L_2$ by

$$\tau(f) = f_*[M_f, L_1] = [M_f \underset{L_1}{\cup} M_f, L_2] \in Wh(L_2).$$

A great deal of formal information about Whitehead groups and torsion can then be deduced from the following facts (exercises for the reader):

Fact 1: If K, L and M are subcomplexes of the complex $K \cup L$, with $M = K \cap L$ and if $K \searrow M$ then $[K \cup L, L] = j_*[K, M]$ where $j : M \to L$ is the inclusion.

Fact 2: If $K \searrow L \searrow M$ and $i : M \to L$ is the inclusion then $[K, M] = [L, M] + (i_*)^{-1}[K, L]$.

However it seems silly to extract this formal information when we cannot do meaningful computations. Conceivably every $Wh(L)$ is 0 and this entire discussion is vacuous. Thus we shall delay drawing out the formal consequences of the preceding discussion until §22–§24, by which time we will have shown that the functor described in (6.2) is naturally equivalent to another functor—one which is highly non-trivial.

Finally we remark that the entire preceding discussion can be modified to apply to (and was developed when the author was investigating) pairs (K, L) of locally finite CW complexes such that there is a proper deformation retraction from K to L. The notion of "elementary collapse" is replaced in the non-compact case by "countable disjoint sequence of finite collapses". For a development of the non-compact theory see [SIEBENMANN] and [FARRELL-WAGONER]. Also the discussion in [ECKMANN-MAUMARY] is valid for locally finite complexes. Finally, the author thinks that [COHEN, §8] is relevant and interesting.

§7. Simplifying a homotopically trivial CW pair

In this section we take a CW pair (K, L) such that $K \rightsquigarrow L$ and simplify it by expanding and collapsing rel L. We start with a lemma which relates the simple-homotopy type of a complex to the attaching maps by which it is constructed.

(7.1) *If* $K_0 = L \cup e_0$ *and* $K_1 = L \cup e_1$ *are CW complexes, where the* e_i $(i = 0, 1)$ *are n-cells with characteristic maps* $\varphi_i: I^n \to K_i$ *such that* $\varphi_0|\partial I^n$ *and* $\varphi_1|\partial I^n$ *are homotopic maps of* ∂I^n *into L, then* $K_0 \nearrow\!\!\searrow K_1$, rel L.

PROOF: We first consider the case where $e_0 \cap e_1 = \varnothing$ and, under this assumption, give the set $L \cup e_0 \cup e_1$ the topology and CW structure which make K_0 and K_1 subcomplexes.

Let $F: \partial I^n \times I \to L$ with $F_i = \varphi_i|\partial I^n$ $(i = 0, 1)$. Give ∂I^n a CW structure and $\partial I^n \times I$ the product structure. Then, by the cellular approximation theorem (3.3) the map $F: (\partial I^n \times I, \partial I^n \times \{0, 1\}) \to (L, L^{n-1})$ is homotopic to a map G such that $G|\partial I^n \times \{0, 1\} = F|\partial I^n \times \{0, 1\}$ and $G(\partial I^n \times I) \subset L^n$. Define $\varphi: \partial(I^n \times I) \to (L \cup e_0 \cup e_1)^n$ by setting

$$\varphi|\partial I^n \times I = G; \qquad \varphi|I^n \times \{i\} = \varphi_i, \qquad i = 0, 1.$$

We now attach an $(n+1)$-cell to $L \cup e_0 \cup e_1$ by φ to get the CW complex

$$K = (L \cup e_0 \cup e_1) \underset{\varphi}{\cup} (I^n \times I).$$

Since $\varphi|I^n \times \{i\}$ is a characteristic map for e_i we have

$$K_0 = L \cup e_0 \nearrow K \searrow L \cup e_1 = K_1, \text{ rel } L.$$

If $e_0 \cap e_1 \neq \varnothing$, construct a CW complex $\hat{K} = L \cup \hat{e}_0$ such that $\hat{e}_0 \cap (e_0 \cup e_1) = \varnothing$ and such that $\hat{e}_0$ has the same attaching map as e_0. Then, by the special case above, $K_0 \nearrow\!\!\searrow \hat{K}_0 \nearrow\!\!\searrow K_1$, rel L. □

As an example, (7.1) may be used to show that the dunce hat D has the same simple-homotopy type as a point. D is usually defined to be a 2-simplex Δ^2 with its edges identified as follows

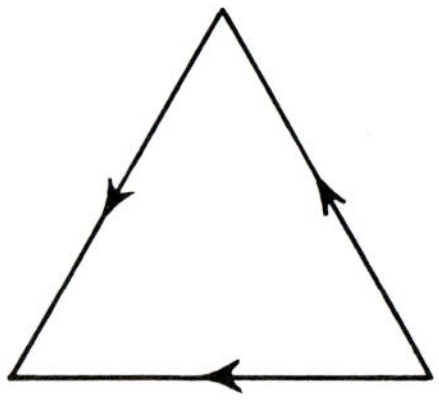

Now D can be thought of as the 1-complex $\partial\Delta^2$ with the 2-cell Δ^2 attached to it by the map $\varphi: \partial\Delta^2 \to \partial\Delta^2$ which takes each edge completely around the circumference once in the indicated direction. Since this map is easily seen to be homotopic to $1_{\partial\Delta^2}$,

$$D = (\partial\Delta^2 \underset{\varphi}{\cup} \Delta^2) \curlywedge (\partial\Delta^2 \underset{1}{\cup} \Delta^2) = \Delta^2 \searrow 0.$$

See [ZEEMAN] for more about the dunce hat.

Before proceeding to the main task of this section we give the following useful consequence of (7.1), albeit one which will not be used in this volume.

(7.2) *Every finite CW complex K has the simple-homotopy type of a finite simplicial complex of the same dimension.*

SKETCH OF PROOF: We shall use the following fact [J. H. C. WHITEHEAD 3 (§15)]

(∗) If J_1 and J_2 are simplicial complexes and $f: J_1 \to J_2$ is a simplicial map then the mapping cylinder M_f is triangulable so that J_1 and J_2 are subcomplexes.

If K is a point the result (7.2) is trivial. Suppose that $K = L \cup e^n$ where e^n is a top dimensional cell with characteristic map $\varphi: I^n \to K$. Set $\varphi_0 = \varphi|\partial I^n$. By induction on the number of cells there is a simple-homotopy equivalence $f: L \to L'$ where L' is a simplicial complex. So, by (5.9),

$$K = L \cup e^n \curlywedge K \underset{f}{\cup} L' = L' \underset{f\varphi_0}{\cup} I^n.$$

Triangulate ∂I^n and let $g: \partial I^n \to L'$ be a simplicial approximation to $f\varphi_0$. Then (7.1) implies that

$$L' \underset{f\varphi_0}{\cup} I^n \curlywedge L' \underset{g}{\cup} I^n.$$

Now $L' \underset{g}{\cup} I^n$ can be subdivided to become a simplicial complex as follows. Consider I^n as $I_0^n \cup (\partial I^n \times I)$ where I_0^n is a concentric cube inside I^n and $\partial I_0^n \equiv \partial I^n \times 0$. Then $|L' \underset{g}{\cup} I^n| = |M_g \cup I_0^n|$. If M_g is triangulated according to (∗) and I_0^n is triangulated as the cone on ∂I_0^n we get a simplicial complex K' with $|K'| = |L' \underset{g}{\cup} I^n|$. It is a fact that

$$L' \underset{g}{\cup} I^n \curlywedge K', \text{ rel } L'.$$

This can be proved by an ad-hoc argument, but it is better for the reader to think of it as coming from the general principle that "subdivision does not change simple-homotopy type", which will be proved in §25. Thus we conclude that $K \nearrow\!\!\searrow K \cup_f L' \nearrow\!\!\searrow L' \cup I^n \nearrow\!\!\searrow K' =$ simplicial complex. □

We now give the basic construction in simplifying a CW pair—that of **trading cells.**

(7.3) *If (K, L) is a pair of connected CW complexes and r is an integer such that*

(a) $\pi_r(K, L) = 0$

(b) $K = L \cup \bigcup_{i=1}^{k_r} e_i^r \cup \bigcup_{i=1}^{k_{r+1}} e_i^{r+1} \cup \ldots \cup \bigcup_{i=1}^{k_n} e_i^n.$

Then $K \nearrow\!\!\searrow M$ rel L where M is a CW complex of the form

$$M = L \cup \bigcup_{i=1}^{k_{r+1}} f_i^{r+1} \cup \bigcup_{i=1}^{k_r + k_{r+2}} f_i^{r+2} \cup \left(\bigcup_{i=1}^{k_{r+3}} f_i^{r+3} \cup \ldots \cup \bigcup_{i=1}^{k_n} f_i^n\right).$$

[Here the e_i^j and f_i^j denote j-cells.]

PROOF: Let $\varphi_i^r: I^r \to K$ be a characteristic map for e_i^r $(i = 1, 2, \ldots, k_r)$. So $\varphi_i^r(\partial I^r) \subset K^{r-1} = L^{r-1}$ and $\varphi_i^r: (I^r, \partial I^r) \to (K, L)$. Since $\pi_r(K, L) = 0$ there is a map $F_i: I^{r+1} \to K$ such that

$$F_i | I^r \times 0 = \varphi_i$$

$$F_i | \partial I^r \times t = \varphi_i | \partial I^r, \qquad 0 \leq t \leq 1$$

$$F_i(I^r \times 1) \subset L.$$

We may assume that, in addition,

$$F_i(\partial I^{r+1}) \subset K^r$$

and

$$F_i(I^{r+1}) \subset K^{r+1}.$$

This is because, if F_i did not have these properties, we could use the cellular approximation theorem as follows. First we would homotop $F_i | \partial I^{r+1}$, relative to $(I^r \times 0) \cup (\partial I^r \times I)$, to a map G_i with $G_i(I^r \times 1) \subset L^r$. By the homotopy extension property, G_i would extend to a map, also called G_i, of I^{r+1} into K. Then $G_i: I^{r+1} \to K$ could be homotoped, relative to ∂I^{r+1}, to $H_i: I^{r+1} \to K^{r+1}$, and H_i would have the desired properties.

Let $P = K \cup_{F_1} I^{r+2} \cup_{F_2} I^{r+2} \cup \ldots \cup_{F_{k_r}} I^{r+2}$ and let $\psi_i: I^{r+2} \to P$ be the identification map determined by the condition that $\psi_i | I^{r+1} \times 0 = F_i$. Recalling that $J^m \equiv \text{Cl}(\partial I^{m+1} - I^m)$, we set

$$E_i^{r+2} = \psi_i(\mathring{I}^{r+2}) \quad \text{and} \quad E_i^{r+1} = \psi_i(\mathring{J}^{r+1}), \quad 1 \leq i \leq k_r$$

Then, by definition of expansion,

$$K \nearrow P = K \cup \bigcup E_i^{r+2}.$$

Consider $P_0 = L \cup \bigcup e_i^r \cup \bigcup E_i^{r+1}$. Thus, when there is a single r-cell and $r = 0$ the situation looks like this:

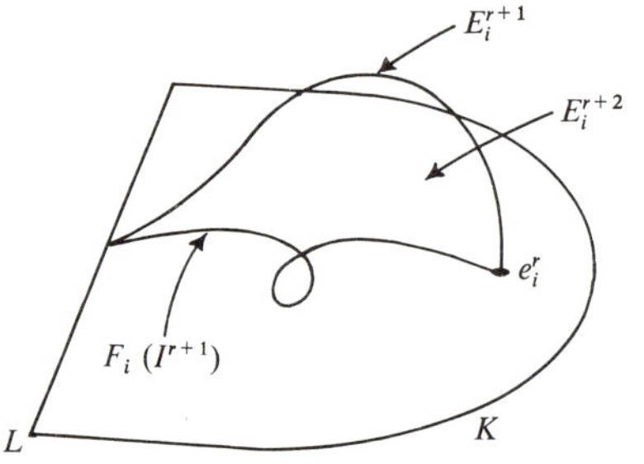

Picture of P

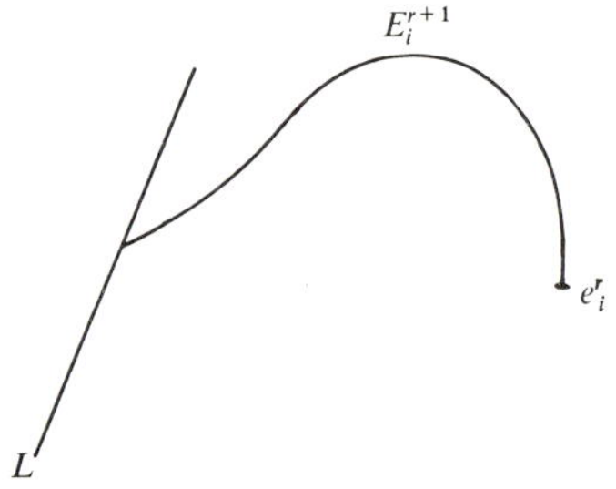

Picture of P_0

Since $\psi_i(\partial J^{r+1}) = F_i(\partial I^{r+1}) \subset K^r$, P_0 is a well-defined subcomplex of P. Also I^r is a face of J^{r+1} such that $\psi_i | I^r = \varphi_i$, a characteristic map for e_i^r. So we have

$$P_0 \searrow L$$

$$P - P_0 = \bigcup e_i^{r+1} \cup \left(\bigcup e_i^{r+2} \cup \bigcup E_i^{r+2}\right) \cup \bigcup e_i^{r+3} \cup \ldots \cup \bigcup e_i^n.$$

Let $g : P_0 \to L$ be a cellular deformation corresponding to this collapse. Applying (5.9), and letting $G : P \to P \underset{g}{\cup} L$ be the map induced by g, we have

$$K \nearrow P \mathrel{\wedge\!\!\!\searrow} P \underset{g}{\cup} L, \text{ rel } L$$

where

$$P \underset{g}{\cup} L = L \cup \bigcup G(e_i^{r+1}) \cup \left[\bigcup G(e_i^{r+2}) \cup \bigcup G(E_i^{r+2})\right] \cup \ldots \cup \bigcup G(e_i^n).$$

The proof is completed by setting $M = P \underset{g}{\cup} L$. □

(7.4) *Suppose that* (K, L) *is a pair of connected CW complexes such that* $K \mathrel{\sim\!\!\!\searrow} L$. *Let* $n = \dim(K - L)$ *and let* $r \geq n-1$ *be an integer. Let* e^0 *be a* 0*-cell of* L. *Then* $K \mathrel{\wedge\!\!\!\searrow} M$, *rel* L, *where*

$$M = L \cup \bigcup_{j=1}^{a} e_j^r \cup \bigcup_{i=1}^{a} e_i^{r+1}$$

and where the e_j^r *and* e_i^{r+1} *have characteristic maps* $\psi_j : I^r \to M$ *and* $\varphi_i : I^{r+1} \to M$ *such that* $\psi_j(\partial I^r) = e^0 = \varphi_i(J^r)$.

Definition: If L is connected, $M \mathrel{\sim\!\!\!\searrow} L$, and (M, L) satisfies the conclusion of (7.4) with $r \geq 2$, then (M, L) is *in simplified form*.

PROOF: Since $K \mathrel{\sim\!\!\!\searrow} L$, $\pi_i(K, L) = 0$ for all i. Thus, by (7.3), we may trade the relative 0-cells of K for 2-cells, then the 1-cells of the new complex for 3-cells, and so on, until we arrive at a complex $\hat{K}$ for which the lowest dimensional cells of $\hat{K} - L$ are r dimensional. Because $r \geq n-1$ there will not be any cells of dimension greater than $r+1$. Thus we may write $\hat{K} = L \cup \bigcup_{j=1}^{a} \hat{e}_j^r \cup \bigcup_{i=1}^{b} \hat{e}_i^{r+1}$. Let the $\hat{e}_j^r$ have characteristic maps $\hat{\psi}_j$.

We claim that, for each j, $\hat{\psi}_j|\partial I^r$ is homotopic in L to the constant map $\partial I^r \to e^0$. For, since $\hat{K} \searrow L$, there is a retraction $R:\hat{K} \to L$. Then $R\hat{\psi}_j: I^r \to L$ and $R\hat{\psi}_j|\partial I^r = \hat{\psi}_j|\partial I^r$ since $\hat{\psi}_j(\partial I^r) \subset \hat{K}^{r-1} \subset L$. Thus $\hat{\psi}_j|\partial I^r$ is null homotopic in L and, L being arc-wise connected, it is homotopic to the constant map at e^0. Therefore by (7.1),

$$L \cup \bigcup \hat{e}_j^r \wedge L \cup \bigcup e_j^r, \text{ rel } L$$

where the e_j^r are trivially attached at e^0. Hence by (5.9)

$$L \cup \bigcup \hat{e}_j^r \cup \bigcup \hat{e}_i^{r+1} \wedge L \cup \bigcup e_j^r \cup \bigcup f_i^{r+1}.$$

Now let the f_i^{r+1} have characteristic maps $\hat{\varphi}_i$. Since J^r is contractible to a point by a homotopy of ∂I^{r+1} the attaching map $\hat{\varphi}_i|\partial I^{r+1}$ is homotopic to a map $\varphi_i: \partial I^{r+1} \to L \cup \bigcup e_j^r$ such that $\varphi_i(J^r) = e^0$. Then, by (7.1) again

$$L \cup \bigcup e_j^r \cup \bigcup f_i^{r+1} \wedge L \cup \bigcup e_j^r \cup \bigcup e_i^{r+1}, \text{ rel } L$$

where the e_i^{r+1} have characteristic maps φ_i such that $\varphi_i(J^r) = e^0$. We call this last complex M.

Finally, to see that the number of r-cells of $M-L$ is equal to the number of $(r+1)$-cells of $M-L$, notice that, by (3.7), these numbers are precisely equal to the ranks of the free (integral) homology modules $H_r(M^r \cup L, L)$ and $H_{r+1}(M, M^r \cup L)$. But since $M \wedge L$, the exact sequence of the triple $(M, M^r \cup L, L)$ contains

$$\to H_{r+1}(M, L) \to H_{r+1}(M, M^r \cup L) \xrightarrow{d} H_r(M^r \cup L, L) \to H_r(M, L) \to$$

where $H_{r+1}(M, L) = H_r(M, L) = 0$. Thus d is an isomorphism and these ranks are equal. □

§8. Matrices and formal deformations

Given a homotopically trivial CW pair, we have shown that it can be transformed into a pair in simplified form. So consider a simplified pair (K, L); $K = L \cup \bigcup e_j^r \cup \bigcup e_i^{r+1}$ where the e_j^r are trivially attached at e^0. If, given r and L, we wish to distinguish one such pair from another, then clearly the crucial information lies in how the cells e_i^{r+1} are attached—i.e., in the maps $\varphi_i|\partial I^{r+1}: \partial I^{r+1} \to L \cup \bigcup e_j^r$, where φ_i is a characteristic map for e_i^{r+1}. Denoting $K_r = L \cup \bigcup e_j^r$, we study these attaching maps in terms of the boundary operator $\partial: \pi_{r+1}(K, K_r; e^0) \to \pi_r(K_r, L; e^0)$ in the homotopy exact sequence of the triple (K, K_r, L). Since, however, *freely* homotopic attaching maps give (7.1) the same result up to simple-homotopy type, we do not wish to be bound to homotopies keeping the base point fixed. To capture this extra degree of freedom formally, we shall think of the homotopy groups not merely as abelian groups, but as modules over $\mathbb{Z}(\pi_1(L, e^0))$. This is done as follows:

Given a pair of connected complexes (P, P_0) and a point $x \in P_0$, it is well-known [SPANIER, §7.3] that $\pi_1 = \pi_1(P_0, x)$ acts on $\pi_n(P, P_0; x)$ by the condition that $[\alpha]\cdot[\varphi] = [\varphi']$, where α and φ represent the elements $[\alpha]$ and $[\varphi]$ of π_1 and $\pi_n(P, P_0; x)$ respectively, and $\varphi':(I^n, I^{n-1}, J^{n-1}) \to (P, P_0, x)$ is homotopic to φ by a homotopy dragging $\varphi(J^{n-1})$ along the loop α^{-1}. This action has the properties that

(0) $[*]\cdot[\varphi] = [\varphi]$, where $[*]$ is the identity element in π_1,

(1) $[\alpha]\cdot([\varphi_1]+[\varphi_2]) = [\alpha]\cdot[\varphi_1]+[\alpha]\cdot[\varphi_2]$,

(2) $([\alpha][\beta])\cdot[\varphi] = [\alpha]\cdot([\beta]\cdot[\varphi])$,

(3) It commutes with all the homomorphisms in the homotopy exact sequence of the pair (P, P_0).

It follows that $\pi_n(P, P_0; x)$ becomes a $\mathbb{Z}\pi_1$-module[7] if we define multiplication by

$$\left(\sum n_j[\alpha_j]\right)[\varphi] = \sum n_j([\alpha_j]\cdot[\varphi]), \quad [\alpha_j] \in \pi_1,\ [\varphi] \in \pi_n(P, P_0; x),$$

and the homotopy exact sequence of $(P, P_0; x)$ becomes an exact sequence of $\mathbb{Z}\pi_1$-modules. In the case of a simplified pair (K, L), the following lemma will be applied to give us the structure of $\pi_{r+1}(K, K_r; e^0)$ and of $\pi_r(K_r, L; e^0)$.

(8.1) *Suppose that (P, P_0) is a CW pair with $P = P_0 \cup \bigcup_{i=1}^{a} e_i^n$, where P_0 is connected. Suppose that $\varphi_i:(I^n, I^{n-1}, J^{n-1}) \to (P, P_0; e^0)$ are characteristic maps for the e_i^n and that either:* a) $n \geq 3$, *or* b) $n = 2$ *and $\varphi_i(\partial I^n) = e^0$ for all i. Then $\pi_n(P, P_0; e^0)$ is a free $\mathbb{Z}\pi_1$-module with basis $[\varphi_1], [\varphi_2], \ldots, [\varphi_a]$.*

PROOF: We claim first that the inclusion map induces an isomorphism $i_\#:\pi_1(P_0, e^0) \to \pi_1(P, e^0)$, For all $n \geq 2$, $i_\#$ is onto because, by the cellular approximation theorem, any map of $(I^1, \partial I^1)$ into (P, e^0) can be homotoped rel ∂I^1 into P_0. Similarly, for all $n \geq 3$, $i_\#$ is one-one, because any homotopy $F:(I^2, \partial I^2) \to (P, P_0)$ between maps F_0 and F_1 can be replaced by a map $G:I^2 \to P_0$ such that $G|\partial I^2 = F|\partial I^2$. Finally, if $n = 2$, $\varphi_i(\partial I^2) = e^0$, by assumption. Let $R:P \to P_0$ be the retraction such that $R(\bigcup e_i^2) = e^0$. Then, if two maps $f, g:(I, \partial I) \to (P_0, e_0)$ are homotopic in P by the homotopy F_t, they are homotopic in P_0 by the homotopy $R \circ F_t$. Hence $i_\#$ is one-one in this case also.

Let $p:\tilde{P} \to P$ be the universal covering of P. Let $\tilde{P}_0 = p^{-1}P_0$. Then $\tilde{P}_0$ is the universal covering space of P_0 with covering map $p|\tilde{P}_0$ (by 3.13). Let G be the group of covering homeomorphisms of $\tilde{P}$. Choose a base point $\tilde{e}^0 \in p^{-1}(e^0)$. For each $i(1 \leq i \leq a)$, let $\tilde{\varphi}_i:(I^n, J^{n-1}) \to (\tilde{P}, \tilde{e}^0)$ cover φ_i. Then (3.15) says that $H_*(\tilde{P}, \tilde{P}_0)$ is a free $\mathbb{Z}(G)$-module with basis $\{\langle\tilde{\varphi}_i\rangle\}$ where $\langle\tilde{\varphi}_i\rangle \equiv (\tilde{\varphi}_i)_*(\omega_n)$, ω_n being a generator of $H_n(I^n, \partial I^n)$. We may first identify

[7] See page 11 for the definition of the group ring $\mathbb{Z}(G)$.

G with $\pi_1(P,e^0)$ (see page 12) and then use the isomorphism $i_\#$ to identify G with $\pi_1(P_0,e^0) = \pi_1$. If $[\alpha] \in \pi_1$, let $g_{[\alpha]}$ be the corresponding covering homeomorphism. Hence $H_*(\tilde{P},\tilde{P}_0)$ is a free $\mathbb{Z}\pi_1$-module with basis $\{\langle\tilde{\varphi}_i\rangle\}$. We complete the proof by demonstrating that $H_n(\tilde{P},\tilde{P}_0)$ is isomorphic to $\pi_n(P,P_0;\, e^0)$ as a $\mathbb{Z}\pi_1$-module, by an isomorphism which takes $\langle\tilde{\varphi}_i\rangle$ onto $[\varphi_i]$ for each i.

To demonstrate this, consider the isomorphism T of $\mathbb{Z}$-modules given by

$$H_n(\tilde{P},\tilde{P}_0) \xrightarrow{h^{-1}} \pi_n(\tilde{P},\tilde{P}_0;\, \tilde{e}^0) \xrightarrow{p_\#} \pi_n(P,P_0;\, e^0)$$

with T denoting the composite (curved arrow from $H_n(\tilde{P},\tilde{P}_0)$ to $\pi_n(P,P_0;\, e^0)$).

Here h is the Hurewicz homomorphism which takes each $[\psi] \in \pi_n(\tilde{P},\tilde{P}_0,\tilde{e}^0)$ onto $\psi_*(\omega_n)$. In fact, applying the Hurewicz theorem [Spanier, p. 397], h is an isomorphism because $\tilde{P}_0$ and $\tilde{P}$ are connected and simply connected and because, by the cellular approximation theorem, $\pi_i(\tilde{P},\tilde{P}_0) = 0$ for $i \le n-1$. Also $p_\#$ is an isomorphism for all $n \ge 1$, by the homotopy lifting property. Thus T is an isomorphism and, clearly, $T(\langle\tilde{\varphi}_i\rangle) = p_\#[\tilde{\varphi}_i] = [p\tilde{\varphi}_i] = [\varphi_i]$. Finally to see that T is a homomorphism *of* $(\mathbb{Z}\pi_1)$*-modules*, it suffices to show that $T(\sum a_i\langle\tilde{\varphi}_i\rangle) = \sum a_i[\varphi_i]$ for all $a_i = \sum_j n_{ij}[\alpha_j] \in \mathbb{Z}\pi_1$. But, by definition of scalar multiplication and our identification of $\mathbb{Z}\pi_1$ with $\mathbb{Z}(G)$,

$$\sum_i a_i\langle\tilde{\varphi}_i\rangle = \sum_i \Big(\sum_j n_{ij}[\alpha_j]\Big)(\tilde{\varphi}_{i*}(\omega_n)) = \sum_{i,j} n_{ij}((g_{[\alpha_j]}\tilde{\varphi}_i)_*(\omega_n)).$$

But $g_{[\alpha_j]}\tilde{\varphi}_i$ is freely homotopic to the map $\tilde{\alpha}_j \cdot g_{[\alpha_j]}\tilde{\varphi}_i$, which is gotten from it by dragging the image of J^{n-1} (namely $g_{[\alpha_j]}(\tilde{e}^0)$) along the path $\tilde{\alpha}_j^{-1}$. Thus, by the homotopy property in homology

$$\begin{aligned}
\sum_i a_i\langle\tilde{\varphi}_i\rangle &= \sum_{i,j} n_{ij}((\tilde{\alpha}_j \cdot g_{[\alpha_j]}\tilde{\varphi}_i)_*(\omega_n)) \\
&\xrightarrow{h^{-1}} \sum_{i,j} n_{i,j}[\tilde{\alpha}_j \cdot g_{[\alpha_j]}\tilde{\varphi}_j] \\
&\xrightarrow{p_\#} \sum_{i,j} n_{i,j}[p \circ (\tilde{\alpha}_j \cdot g_{[\alpha_j]}\tilde{\varphi}_i] \\
&= \sum_{i,j} n_{i,j}([\alpha_j] \cdot [\varphi_i]) \\
&= \sum_i \Big(\sum_j n_{i,j}[\alpha_j]\Big)[\varphi_i] \\
&= \sum_i a_i[\varphi_i]. \qquad \square
\end{aligned}$$

Suppose now that (K, L) is in simplified form, where

$$K = L \cup \bigcup_{j=1}^{a} e_j^r \cup \bigcup_{i=1}^{a} e_i^{r+1}.$$

Let $\{\varphi_i\}$ and $\{\psi_j\}$ be characteristic maps for the e_i^{r+1} and e_j^r respectively. Then by the preceding lemma, $\{[\varphi_i]\}$ and $\{[\psi_j]\}$ are bases for the $\mathbb{Z}\pi_1$-modules $\pi_{r+1}(K, K_r)$ and $\pi_r(K_r, L)$, where $K_r = L \cup \bigcup e_j^r$. We define *the matrix of*

(K, L) with respect to the characteristic maps $\{\varphi_i\}$ and $\{\psi_j\}$ to be the $(a \times a)$ $\mathbb{Z}\pi_1$-matrix (a_{ij}), given by $\partial[\varphi_i] = \sum a_{ij}[\psi_j]$ where $\partial: \pi_{r+1}(K, K_r) \to \pi_r(K_r, L)$ is the usual boundary operator. Notice that this matrix must be non-singular (i.e., have a 2-sided inverse). For $\pi_{r+1}(K, L) = \pi_r(K, L) = 0$, since $K \searrow\!\!\nearrow L$; so, by exactness of the homotopy sequence, ∂ is an isomorphism.

The simplest example of a pair in simplified form occurs when the characteristic maps φ_i, ψ_j satisfy $\varphi_i(J^r) = e^0$ and $\varphi_i|I^r = \psi_i$. In this case we have, algebraically, that the matrix of (K, L) with respect to the given bases is the $a \times a$ identity matrix and, geometrically, that $K \searrow L$. (In fact $K = L \cup$ [wedge product of balls].) More generally, when the matrix is right we can cancel cells as follows:

(8.2) *If (K, L) is a simplified pair and if the matrix of (K, L) with respect to some choice of characteristic maps $\{\varphi_i\}$, $\{\psi_j\}$ is the identity, then $K \searrow\!\!\nearrow L$,* rel *L*.

PROOF: Consider the characteristic maps $\varphi_1:(I^{r+1}, I^r, J^r) \to (K, K_r, e^0)$ and $\psi_1:(I^r, \partial I^r) \to (K_r, e^0)$. By hypothesis, $[\psi_1] = \partial[\varphi_1] \equiv [\varphi_1|I^r] \in \pi_r(K_r, L; e^0)$. Thus there is a homotopy $h_t:(I^r, I^{r-1}, J^{r-1}) \to (K_r, L; e^0)$ such that $h_0 = \varphi_1|I^r$ and $h_1 = \psi_1$. By the homotopy extension theorem (3.1) we may extend $h_t|\partial I^r$ to a homotopy $g_t: J^r \to L$ such that $g_0|J^r = \varphi_1|J^r$. Combining h_t and g_t we have a homotopy $H_t:(\partial I^{r+1}, I^r, J^r) \to (K_r, K_r, L)$ with $H_0 = \varphi_1|\partial I^{r+1}$ and $H_1|I^r = \psi_1$. By the cellular approximation theorem H_1 can be homotoped, rel I^r, to $\hat{\varphi}_1$ where $\hat{\varphi}_1(J^r) \subset L^r$. If we attach an $(r+1)$-cell $\hat{e}_1^{r+1}$ to K_r by $\hat{\varphi}_1: \partial I^{r+1} \to K_r$ then, by (7.1) we have

$$K = L \cup \bigcup_j e_j^r \cup \bigcup_i e_i^{r+1} \searrow\!\!\nearrow \left(L \cup \bigcup_j e_j^r \cup \bigcup_{i>1} e_i^{r+1}\right) \cup \hat{e}_1^{r+1}, \text{ rel } L$$

$$\searrow L \cup \bigcup_{j>1} e_j^r \cup \bigcup_{i>1} e_1^{r+1} \equiv K'.$$

The last collapse takes place because $\hat{\varphi}_1|I^r = \psi_1$.

Finally, the matrix of (K', L) with respect to the remaining characteristic maps is the identity matrix with one fewer row and column. For suppose that $\partial': \pi_{r+1}(K', K_r') \to \pi_r(K_r', L)$, that $i': K' \subset K$ and that $\varphi_i = i'\varphi_i'$, $\psi_j = i'\psi_j'$. If $\partial'[\varphi_i'] = \sum a_{ij}'[\psi_j']$ then $[\psi_i] = \partial[\varphi_i] = i'_\# \partial'[\varphi_i'] = i'_\# \sum a_{ij}'[\psi_j'] = \sum a_{ij}'[\psi_j]$. So $a_{ij}' = \delta_{ij}$. Thus we may proceed by induction on the number of cells of $K-L$. □

Exercise: Go through the preceding proof in the example where $L = e^0 \cup e_0^2$ (the 2-sphere), $K_r = L \cup e^2$ and the sole 3-cell is attached by $\varphi: \partial I^3 \to L \cup e^2$ such that

$$\varphi(J^2 \cup \{(\tfrac{1}{2}, y, 0) \mid 0 \le y \le 1\}) = e^0$$

$$\varphi|\{(x, y, 0) \mid 0 \le x \le \tfrac{1}{2}, 0 \le y \le 1\} = \text{characteristic map for } e_0^2$$

$$\varphi|\{(x, y, 0) | \tfrac{1}{2} \le x \le 1, 0 \le y \le 1\} = \text{characteristic map for } e^2.$$

If the matrix of the simplified pair (K, L) is not the identity we might nevertheless be able to expand and collapse to get a new pair (M, L) whose matrix is the identity. The following lemma shows that certain algebraic changes of the matrix of a given pair can be realized by expanding and collapsing.

(8.3) *Assume that the pair (K, L) is in simplified form and has matrix (a_{ij}) with respect to some set of characteristic maps. Suppose further that the matrix (a_{ij}) can be transformed to the matrix (b_{ij}) by one of the following operations*

I. $R_i \to \pm\alpha R_i \quad (\alpha \in \pi_1 \subset \mathbb{Z}\pi_1)$

(Multiply the i'th row on the left by plus or minus an element of the group)

II. $R_k \to R_k + \rho R_i \quad (\rho \in \mathbb{Z}\pi_1)$

(Add a left group-ring multiple of one row to another)

III. $(a_{ij}) \to \begin{pmatrix} a_{ij} & 0 \\ 0 & I_q \end{pmatrix}$

(Expand by adding a corner identity matrix)

Then there is a simplified pair (M, L) such that $K \nearrow\!\!\!\searrow M$ rel L and a set of characteristic maps with respect to which (M, L) has the matrix (b_{ij}).

PROOF: Suppose, as usual, that $K = L \cup \bigcup e_j^r \cup \bigcup e_i^{r+1}$, and denote the given characteristic maps for the r and $(r+1)$ cells by $\{\psi_j\}$ and $\{\varphi_i\}$ respectively. For notational simplicity we consider I. when $R_1 \to \pm\alpha R_1$ and II. when $R_1 \to R_1 + \rho R_2$.

To realize $R_1 \to -R_1$, set $M = K$ and introduce the new characteristic map $\hat{\varphi}_1$ to replace φ_1, where $\hat{\varphi}_1 = \varphi_1 \circ R$ and $R: I^{r+1} \to I^{r+1}$ by $R(x_1, x_2, \ldots, x_{r+1}) = (1 - x_1, x_2, \ldots, x_{r+1})$. Clearly $[\hat{\varphi}_1] = -[\varphi_1]$, so $\partial[\hat{\varphi}_1] = -\partial[\varphi_1] = -\sum a_{ij}[\psi_j]$. To realize $R_1 \to \alpha R_1$, let $f:(I^r, \partial I^r) \to (K_r, e^0)$ represent $\alpha \cdot [\varphi_1 | I^r] \in \pi_r(K_r, L)$. Extend f trivially to ∂I^{r+1}. Set

$$M = L \cup \bigcup_j e_j^r \cup \bigcup_{i>1} e_i^{r+1} \cup \hat{e}_1^{r+1}$$

where $\hat{e}_1^{r+1}$ has characteristic map $\hat{\varphi}_1$ with $\hat{\varphi}_1 | \partial I^{r+1} = f$. Clearly $\partial[\hat{\varphi}_1] = \alpha \cdot \partial[\varphi_1]$. But $\hat{\varphi}_1 | \partial I^{r+1}$ is freely homotopic in K_r to $\varphi_1 | \partial I^{r+1}$. Thus, by (7.1), $K \nearrow\!\!\!\searrow M$ rel L.

To realize the operation $R_1 \to R_1 + \rho R_2$, let $\varphi:(I^{r+1}, I^r, J^r) \to (K, K_r, e^0)$ be the canonical representative of $[\varphi_1] + [\varphi_2']$, where φ_2' represents $\rho \cdot [\varphi_2]$

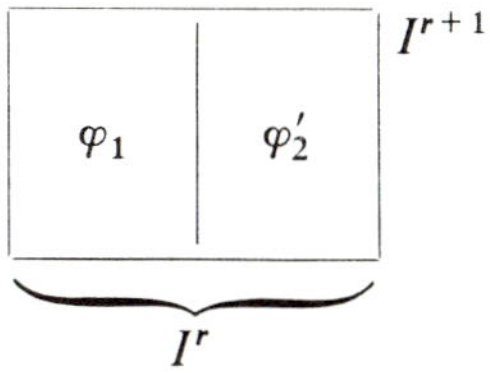

Then $\partial[\varphi] = \partial[\varphi_1] + \rho \cdot \partial[\varphi_2] = \sum_j (a_{1j} + \rho a_{2j})[\psi_j]$. Notice that $\varphi(\partial I^{r+1}) \subset K_r$

$\subset K_r \cup e_2^{r+1}$. Also $\varphi|\partial I^{r+1}$ is homotopic in $K_r \cup e_2^{r+1}$ to $\varphi_1|\partial I^{r+1}$ because $\varphi_2'|I^r$ is homotopic (rel ∂I^r) in $K_r \cup e_2^{r+1}$ to the constant map at e^0. (In fact φ_2' is the homotopy!) Therefore we may attach a new cell with characteristic map $\hat{\varphi}_1$ such that $\hat{\varphi}_1|\partial I^{r+1} = \varphi|\partial I^{r+1}$ and thus construct a new complex M with the desired matrix such that

$$K = [K_r \cup \bigcup_i e_i^{r+1}] \nearrow\!\!\searrow [K_r \cup \bigcup_{i>1} e_i^{r+1} \underset{\varphi|\partial I^{r+1}}{\cup} I^{r+1}] = M, \text{ rel } K_r \cup \bigcup_{i>1} e_i^{r+1}.$$

Finally, a matrix operation of type III may be realized by elementary expansions of K. □

(8.4) *Suppose that (K, L) is a pair in simplified form which has matrix A with respect to some set of characteristic maps. Suppose further that A can be transformed to an identity matrix I_q by operations of type* (I)–(V), *where* (I), (II) *and* (III) *are as in* (8.3) *and* (IV) *and* (V) *are the analogous column operations.*

IV. $C_j \to \pm C_j \cdot \alpha \qquad (\alpha \in \pi_1 \subset Z\pi_1)$

V. $C_k \to C_k + C_i\rho \qquad (\rho \in Z\pi_1)$

Then $K \nearrow\!\!\searrow L$ rel L.

PROOF: Suppose that $(a_{ij}) \to I_q$ by these five types of operations. Obviously the type III operations may all be done first. Then, as is well known, operations of type I and II (IV and V) correspond to left (right) multiplication by elementary matrices. [By an elementary matrix we mean either a diagonal matrix with all 1's except a single $\pm\alpha$ $(\alpha \in \pi_1)$ on the diagonal, or a matrix which has all ones on the diagonal and a single non-zero entry $a_{ij} = \rho(\rho \in Z\pi_1)$ off of the diagonal.] Thus we have

$$I_q = B\begin{pmatrix} A & 0 \\ 0 & I \end{pmatrix} C, \quad [B, C \text{ products of elementary matrices}]$$

$$C^{-1} = B\begin{pmatrix} A & 0 \\ 0 & I \end{pmatrix}$$

$$I_q = CB\begin{pmatrix} A & 0 \\ 0 & I \end{pmatrix} \quad [CB \text{ a product of elementary matrices}]$$

So A can be transformed to the identity by operations (I), (II), (III) only. Hence by (8.2) and (8.3), $K \nearrow\!\!\searrow L$ rel L. □

We come now to our first major theorem.

(8.5) *If (K, L) is a CW pair such that K and L are 1-connected and $K \searrow\!\!\!\swarrow L$ then $K \nearrow\!\!\searrow L$ rel L.*

PROOF: By (7.4) $K \nearrow\!\!\searrow J$ rel L, where (J, L) is in simplified form. Let A be the matrix of (J, L) with respect to some set of characteristic maps. Then since $\pi_1 L = \{1\}$, $Z\pi_1 = Z$. Thus A is a non-singular matrix with integral coefficients. It is well-known that such a matrix can be transformed to the identity matrix by operations of types (I), (II), (IV) and (V). Therefore, by (8.4), $J \nearrow\!\!\searrow L$ rel L. □

The proof in (8.5) depends on the fact that $\mathbb{Z}$ is a ring over which non-singular matrices can be transformed to the identity. The next lemma shows that the algebra is not always so simple.

If G is a group then a *unit* in $\mathbb{Z}(G)$ is an element with a two-sided multiplicative inverse. The elements of the group $\pm G = \{g|g \in G\} \cup \{-g|g \in G\}$ are called the *trivial* units, and all others are called the *non-trivial* units of $\mathbb{Z}(G)$.

(8.6) *A. Suppose that G is an abelian group such that $\mathbb{Z}(G)$ has non-trivial units. Then there is a non-singular $\mathbb{Z}(G)$ matrix A which cannot be transformed to an identity matrix by any finite sequence of the operations* (I)–(V).

B. The group $G = \mathbb{Z}_5$ is an abelian group such that $\mathbb{Z}(\mathbb{Z}_5)$ has non-trivial units. (Infinitely many such groups will be given in (11.3).)

PROOF: Let a be a non-trivial unit of $\mathbb{Z}(G)$ and let $A = (a)$ be the one-by-one matrix with a as entry. Since G is abelian, $\mathbb{Z}(G)$ is a commutative ring. Thus the determinant operation gives a well-defined map from the square matrices over $\mathbb{Z}(G)$ to $\mathbb{Z}(G)$ which satisfies the usual properties of determinants. The operations (II), (III) and (V) transform any given matrix into another matrix with the same determinant. Operations (I) and (IV) multiply the determinant by a trivial unit. Thus if B is a matrix into which A can be transformed, $\det B = g\cdot(\det A) = ga$ for some trivial unit g. Therefore $\det B$ is a non-trivial unit and, in particular, B cannot be an identity matrix.

To see that $\mathbb{Z}(\mathbb{Z}_5)$ has non-trivial units, let $\mathbb{Z}_5 = \{1, t, t^2, t^3, t^4\}$. Then $a = 1-t+t^2$ is a non-trivial unit since $(1-t+t^2)(t+t^2-t^4) = 1$. □

The next lemma shows that the algebraic difficulties illustrated in (8.6) can, in fact, always be realized geometrically.

(8.7) *If G is a group which can be finitely presented and A is a non-singular $\mathbb{Z}(G)$ matrix then*

(1) *There is a connected CW complex L with $\pi_1(L, e^0) = G$.*

(2) *For any connected complex L with $\pi_1(L, e^0) = G$, there is a CW pair (K, L) in simplified form such that the matrix of (K, L) with respect to some set of characteristic maps is precisely A.*

PROOF: Suppose that G is given by generators $x_1, \ldots, x_m$ and relations $R_i(x_1, \ldots, x_m) = 1$, $(i = 1, 2, \ldots, n)$. Let $L^1 = e^0 \cup (e_1^1 \cup \ldots \cup e_m^1)$, a wedge product of circles, and let x_j be the element of the free group $\pi_1(L^1, e^0)$ represented by a characteristic map for e_j^1. Let $\varphi_i : \partial I^2 \to L^1$ represent the element $R_i(x_1, \ldots, x_m)$. Finally let $L = L^1 \cup_{\varphi_1} I^2 \cup_{\varphi_2} \ldots \cup_{\varphi_n} I^2$. By successive applications of VAN KAMPEN's theorem, $\pi_1(L, e^0)$ is precisely G.

In proving (2), write $A = (a_{ij})$, a $p \times p$ matrix. Let $K_2 = L \cup e_1^2 \cup \ldots \cup e_p^2$ where the e_j^2 have characteristic map ψ_j with $\psi_j(\partial I^2) = e^0$. As usual $[\psi_j]$ denotes the element of $\pi_2(K_2, L)$ represented by $\psi_j : (I^2, I^1, J^1) \to (K_2, L, e^0)$.

Let $\langle\psi_j\rangle$ denote the element of $\pi_2(K_2, e^0)$ represented by ψ_j and let $\alpha:(K_2, e^0, e^0) \subsetneq (K_2, L, e^0)$. Clearly $\alpha_\#\langle\psi_j\rangle = [\psi_j]$. Let $f_i:(I^2, \partial I^2) \to (K_2, e^0)$ represent $\sum_j a_{ij}\langle\psi_j\rangle$. Finally, attach 3-cells to K_2 to get $K = K_2 \cup e_1^3 \cup \ldots \cup e_p^3$ where the e_i^3 have characteristic maps $\varphi_i:(I^3, I^2, J^2) \to (K, K_2, e^0)$ with $\varphi_i|I^2 = \alpha \circ f_i$. Then $\partial[\varphi_i] = [\varphi_i|I^2] = [\alpha \circ f_i] = \alpha_\# (\sum_j a_{ij}\langle\psi_j\rangle) = \sum a_{ij}[\psi_j]$. Thus we have constructed a pair (K, L) with $K-L = \bigcup e_j^2 \cup \bigcup e_i^3$ such that the boundary operator $\partial:\pi_3(K, K_2, e^0) \to \pi_2(K_2, L, e^0)$ has matrix A.

It remains only to show that $K \searrow L$. It suffices by (3.2) to show that $\pi_n(K, L) = 0$ for $n \leq 3$. For $n \leq 1$, this is clear from the cellular approximation theorem and the connectivity of K and L. For $n = 2, 3$ we use the fact that ∂ is an isomorphism because A was assumed non-singular. Thus, for $n = 2$, we have the sequence

$$\pi_3(K, K_2) \xrightarrow[\cong]{\partial} \pi_2(K_2, L) \to \pi_2(K, L) \to \underbrace{\pi_2(K, K_2)}_{0}$$

and by exactness it follows that $\pi_2(K, L) = 0$. (Here $\pi_2(K, K_2) = 0$ because $K-K_2$ is the union of 3-cells.) Finally note that $\pi_3(K, L) \cong \pi_3(\tilde{K}, \tilde{L}) \cong H_3(\tilde{K}, \tilde{L})$, the last isomorphism coming from the Hurewicz theorem which applies because $0 = \pi_i(K, L) \cong \pi_i(\tilde{K}, \tilde{L})$ for $i = 1, 2$ and because $\tilde{L}$ is 1-connected by (3.13). To see that $H_3(\tilde{K}, \tilde{L}) = 0$ consider the commutative diagram

$$\begin{array}{ccccccc}
0 = H_3(\tilde{K}_2, \tilde{L}) \to H_3(\tilde{K}, \tilde{L}) \to & H_3(\tilde{K}, \tilde{K}_2) & \xrightarrow{\tilde{\partial}} & H_2(\tilde{K}_2, \tilde{L}) \\
& \text{Hurewicz} \downarrow \cong & & \text{Hurewicz} \downarrow \cong \\
& \pi_3(\tilde{K}, \tilde{K}_2) & \longrightarrow & \pi_2(\tilde{K}_2, \tilde{L}) \\
& \downarrow \cong & & \downarrow \cong \\
& \pi_3(K, K_2) & \xrightarrow[\cong]{\partial} & \pi_2(K_2, L)
\end{array}$$

Clearly $\tilde{\partial}$ is an isomorphism so that, by exactness of the top line, $H_3(\tilde{K}, \tilde{L}) = 0$. Hence $\pi_3(K, L) = 0$. $\square$

Summarizing the situation: It has been shown that in certain cases a homotopically trivial pair (K, L) must have $K \searrow L$ rel L. This occurs (8.5) when $\pi_1 L = 0$ or, more generally, (8.4), when all non-singular matrices over $\mathbb{Z}(\pi_1 L)$ can be transformed into identity matrices. Thus, by §4 and §5, the concepts of homotopy equivalence and simple-homotopy equivalence coincide among CW complexes with sufficiently nice fundamental groups. On the other hand we have exhibited ((8.6) and (8.7)) simplified pairs with matrices which cannot be transformed to an identity matrix. We must ask now whether these matrices—or better, their equivalence classes under operations (I)–(III)—are intrinsic to the problem or whether they are merely artifacts. Starting with a pair (K, L) such that $K \nearrow\!\!\searrow L$, does the equivalence class of the matrix which appears when (K, L) is expanded and collapsed to a pair in simplified form depend on the particular choice of formal deformation?

Two observations are crucial. First, the equivalence classes of non-singular matrices form a group, the Whitehead group of $\pi_1 L$—written $Wh(\pi_1 L)$. (This will be proved in the next chapter.) Second, if $K \nearrow\!\!\searrow J$ rel L where (J, L) is in simplified form then it is implicit in the proof of (8.1) that the matrix of (J, L) is the matrix of the boundary operator

$$H_{r+1}(\tilde{J}, \tilde{J}_r) \to H_r(\tilde{J}_r, \tilde{L}).$$

By definition of the cellular chain complex (page 7), this is the boundary operator in $C(\tilde{J}, \tilde{L})$ where $C(\tilde{J}, \tilde{L})$ is the chain complex

$$0 \to C_{r+1}(\tilde{J}, \tilde{L}) \xrightarrow{\partial} C_r(\tilde{J}, \tilde{L}) \to 0.$$

Since $\tilde{J} \nearrow\!\!\searrow \tilde{L}$, $C(\tilde{J}, \tilde{L})$ is an acyclic $\mathbb{Z}(\pi_1 L)$-complex. Thus to an acyclic $\mathbb{Z}(\pi_1 L)$-complex we have associated an element of $Wh(\pi_1 L)$. We would like to show that the element which is thus determined by $C(\tilde{J}, \tilde{L})$ is pre-determined by $C(\tilde{K}, \tilde{L})$ and, indeed, by (K, L).

At this point a more sophisticated and algebraic approach is necessary. The next chapter will consist of a purely algebraic study of acyclic chain complexes, of the Whitehead groups of groups, and of the rich tapestry which can be woven from these strands.

Chapter III

Algebra

§9. Algebraic conventions

Rings and modules:

Throughout Chapter III, R will denote a ring with unity satisfying:

(∗) *If M is any finitely generated free module over R then any two bases of M have the same cardinality.*

All modules will be assumed to be finitely generated left modules—i.e., when multiplying, ring elements are written to the left of module elements.

It is an elementary exercise that a finitely generated free module has only finite bases. Thus, by these conventions, a "free R-module" always means an R-module with finite bases, any two of which have the same cardinality.

It is well known that division rings satisfy (∗). More generally we have:

(9.1) *The condition (∗) is satisfied by the ring R if there is a division ring D and a non-zero ring homomorphism $f: R \to D$.*

PROOF: By considering the matrices which occur in changing bases, one can see that (∗) is satisfied by a given ring if and only if every matrix A, with entries in R, for which there is a matrix B with $AB = I_m$ and $BA = I_n$ is square (i.e. has $m = n$).

Let f_* be the induced map taking matrices over R into matrices over D given by $f_*((a_{ij})) = (f(a_{ij}))$. Since f is a ring homomorphism $f_*(AB) = f_*(A)f_*(B)$ for all A, B. Now suppose that A and B are arbitrary matrices such that $AB = I_m$ and $BA = I_n$. Because $f(1)$ is a unit, it follows that $f(1) = 1$. So $f_*(I_q) = I_q$ for all q. Thus $f_*(A)f_*(B) = I_m$ and $f_*(B)f_*(A) = I_n$. Hence, since D is a division ring, $f_*(A)$ is square, implying that A is square. Therefore R satisfies (∗). □

(9.2) *If G is a group then $\mathbb{Z}(G)$ satisfies (∗).*

PROOF: The augmentation map $A: \mathbb{Z}(G) \to$ (rationals) given by $A(\sum_i n_i g_i) = \sum_i n_i$, is a non-zero ring homomorphism. Apply (9.1). □

Matrices:

If $f: M_1 \to M_2$ is a module homomorphism where M_1 and M_2 have ordered bases $x = \{x_1, \ldots, x_p\}$ and $y = \{y_1, \ldots, y_q\}$, respectively, then $\langle f \rangle_{x,y}$ denotes the matrix (a_{ij}) where $f(x_i) = \sum_j a_{ij} y_j$. Thus each row of $\langle f \rangle_{x,y}$ gives the image of a basis element of x. When the bases are clear from the context we simply write $\langle f\rangle$ to denote this matrix. When the meaning is extraordinarily unambiguous we may sometimes write f instead of $\langle f\rangle$.

Beware of the fact that these conventions lead to

$$\langle f_2 \circ f_1\rangle = \langle f_1\rangle\langle f_2\rangle.$$

If $x = \{x_1, \ldots, x_p\}$ and $y = \{y_1, \ldots, y_p\}$ are two ordered bases of the same module, M, then $\langle x/y\rangle$ denotes the non-singular matrix (a_{ij}) where $x_i = \sum_j a_{ij}y_j$. If x, y, z are bases of M then $\langle x/z\rangle = \langle x/y\rangle\langle y/z\rangle$.

Suppose that $f\colon M_1 \to M_2$ is a module homomorphism, and that x and x' are bases for M_1, and y and y' are bases for M_2. Then

$$\langle f\rangle_{x',y'} = \langle x'/x\rangle\langle f\rangle_{x,y}\langle y/y'\rangle,$$

a simple formula to remember.

The fact that a matrix can represent either a map or a change of basis has the following expression in this notation. If $x = \{x_1, \ldots, x_p\}$ and $y = \{y_1, \ldots, y_p\}$ are two bases for the module M and if $f\colon M \to M$ is the isomorphism given by $f(y_i) = x_i$ for all i, then $\langle f\rangle_{y,y} = \langle x/y\rangle$.

Direct sums:

If $f\colon A \to C$ and $g\colon B \to D$ then $f \oplus g\colon A \oplus B \to C \oplus D$ is defined by $(f \oplus g)(a, b) = (f(a), g(b))$.

§10. The groups $K_G(R)$

The group of non-singular $n \times n$ matrices (i.e., matrices which have a two-sided inverse) over the ring R is denoted by $GL(n, R)$. There is a natural injection of $GL(n, R)$ into $GL(n+1, R)$ given by

$$A \mapsto \begin{pmatrix} A & \\ & 1 \end{pmatrix}.$$

Using this, *the infinite general linear group of* R is defined as the direct limit $GL(R) = \varinjlim GL(n, R)$. (Alternatively, $GL(R)$ may be thought of as the group consisting of all infinite non-singular matrices which are eventually the identity.) For notational convenience we shall identify each $A \in GL(n, R)$ with its image in $GL(R)$.

Let $E^n_{i,j}$ $(i \neq j)$ be the $n \times n$ matrix with all entries 0 except for a 1 (unity element in R) in the (i, j)-spot. An *elementary matrix* is a matrix of the form $(I_n + aE^n_{i,j})$ for some $a \in R$. We let $E(R)$ denote the subgroup of $GL(R)$ generated by the elementary matrices. Elements of $E(R)$ will be denoted by E, E_1, E_2, etc.

In order to study $GL(R)/E(R)$, define an equivalence relation on $GL(R)$ by:

$$A \sim B \Leftrightarrow \text{there are elements } E_1, E_2 \in E(R) \text{ such that } A = E_1BE_2.$$

We will shortly prove (10.2) that $E(R)$ is normal, so that this is just the relationship of belonging to the same coset of $E(R)$. In the meantime it is clear that $A \sim B$ iff A can be gotten from B by a finite sequence of operations which consist of adding a left multiple of one row to another, or a right multiple of one column to another. More generally, instead of using rows, if P_1 and P_2 are disjoint $p \times n$ and $q \times n$ submatrices of the non-singular $n \times n$ matrix A and if X is a $p \times q$ matrix, the following hold

$$I_R: \quad A = \begin{pmatrix} \hline P_1 \\ \hline P_2 \\ \hline \end{pmatrix} \sim B = \begin{pmatrix} \hline P_1 + XP_2 \\ \hline P_2 \\ \hline \end{pmatrix}$$

II_R: If $p = q$,

$$A = \begin{pmatrix} \hline P_1 \\ \hline P_2 \\ \hline \end{pmatrix} \sim B = \begin{pmatrix} \hline P_2 \\ \hline -P_1 \\ \hline \end{pmatrix}$$

I_R is immediate from the definition of matrix multiplication. II_R follows from I_R by the sequence

$$\begin{pmatrix} \hline P_1 \\ \hline P_2 \\ \hline \end{pmatrix} \to \begin{pmatrix} \hline P_1 + P_2 \\ \hline P_2 \\ \hline \end{pmatrix} \to \begin{pmatrix} \hline P_1 + P_2 \\ \hline -P_1 \\ \hline \end{pmatrix} \to \begin{pmatrix} \hline P_2 \\ \hline -P_1 \\ \hline \end{pmatrix}$$

The corresponding operations on columns give rise to analogous equivalences which we call I_C and II_C.

(10.1) *If A, B are elements of $GL(R)$ then $AB \sim BA$.*

PROOF: For sufficiently large n we may assume that A and B are both $n \times n$ matrices. Then

$$AB = \begin{pmatrix} AB & 0 \\ 0 & I_n \end{pmatrix} \sim \begin{pmatrix} AB & A \\ 0 & I_n \end{pmatrix} \sim \begin{pmatrix} 0 & A \\ -B & I_n \end{pmatrix} \sim \begin{pmatrix} 0 & A \\ -B & 0 \end{pmatrix}$$

Similarly $BA \sim \begin{pmatrix} 0 & B \\ -A & 0 \end{pmatrix}$

Finally, using II_C and II_R

$$\begin{pmatrix} 0 & A \\ -B & 0 \end{pmatrix} \sim \begin{pmatrix} A & 0 \\ 0 & B \end{pmatrix} \sim \begin{pmatrix} 0 & B \\ -A & 0 \end{pmatrix}. \quad \square$$

(10.2) *$E(R)$ is the commutator subgroup of $GL(R)$.*

PROOF: If $E \in E(R)$ and $X \in GL(R)$, then $(XE)X^{-1} \sim X^{-1}(XE)$ by (10.1), so $XEX^{-1} = E_1EE_2 \in E(R)$. Given a commutator $ABA^{-1}B^{-1}$ we apply this with $X = BA$ to get

$$(AB)(A^{-1}B^{-1}) \overset{(10.1)}{=} [E_1(BA)E_2](BA)^{-1} = E_1[(BA)E_2(BA)^{-1}] \in E(R).$$

Hence the commutator subgroup is contained in $E(R)$.

Conversely, a typical generator of $E(R)$ is of the form $(I_n + aE^n_{i,k})$. Noticing that $(I_n + aE^n_{i,j})^{-1} = (I^n - aE^n_{i,j})$, we can see that this generator is a commutator because

$$(I_n + aE^n_{i,k}) = (I_n + aE^n_{i,j})(I_n + E^n_{j,k})(I_n - aE^n_{i,j})(I_n - E^n_{j,k}). \quad \square$$

From elementary algebra this gives immediately

(10.3) If H is a subgroup of $GL(R)$ containing $E(R)$ then H is a normal subgroup and $GL(R)/H$ is abelian. $\square$

Suppose that G is a subgroup of the group of units of R. Let E_G be the group generated by $E(R)$ and all matrices of the form

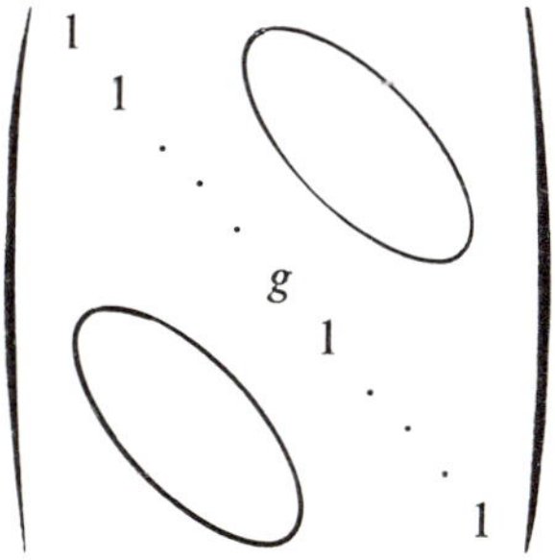

where $g \in G$. Then we define

$$K_G(R) \equiv \frac{GL(R)}{E_G}.$$

By (10.3) this is an abelian group. We denote the quotient map by $\tau: GL(R) \to K_G(R)$ and we call $\tau(A)$ the *torsion of the matrix A*. Since $K_G(R)$ is abelian and will be written additively, we have $\tau(AB) = \tau(A) + \tau(B)$.

Examples of $K_G(R)$ for the most popular choices of G:

1. $K_1(R) = \dfrac{GL(R)}{E(R)}, \qquad G = \{1\}$
2. $\bar{K}_1(R) = K_G(R), \qquad G = \{+1, -1\}$
3. $Wh(G) = K_T(\mathbb{Z}(G))$, where G is a given group, $R = \mathbb{Z}(G)$ and $T = G \cup (-G)$ is the group of trivial units of $\mathbb{Z}(G)$.

$\bar{K}_1(R)$ has the advantage, as does any $K_G(R)$ with $-1 \in G$, that multiplying a row (or column) by (-1) does not change the torsion of a matrix, and so, by II_R (II_C) neither does the interchange of two rows (or columns). The *Whitehead group of G*, $Wh(G)$, is the most important example for our purposes.

If G and G' are subgroups of the units of R and R' respectively then any ring homomorphism $f: R \to R'$ such that $f(G) \subset G'$ induces a group homomorphism $f_*: K_G(R) \to K_{G'}(R')$ given by

$$f_*\tau((a_{ij})) = \tau((f(a_{ij}))).$$

f_* is well-defined because, if $(a_{ij}) \in E_G$, then $(f(a_{ij})) \in E_{G'}$. Thus we have a covariant functor

$$\left\{\begin{array}{l} \text{pairs } (R, G) \\ \\ \text{ring homomorphisms} \\ f: R \to R' \quad \text{with} \quad f(G) \subset G' \end{array}\right\} \to \left\{\begin{array}{l} \text{abelian groups } K_G(R) \\ \\ \text{group homomorphisms} \\ f_*: K_G(R) \to K_{G'}(R') \end{array}\right\}$$

This then gives rise to a covariant functor from the category of groups, and group homomorphisms to the category of abelian groups and group homomorphisms given by

$$G \mapsto Wh(G)$$

$$(f: G \to G') \mapsto (f_*: Wh(G) \to Wh(G'))$$

where f first induces the ring homomorphism $\mathbb{Z}(G) \to \mathbb{Z}(G')$ given by $\sum n_i g_i \to \sum n_i f(g_i)$, and this in turn induces f_* as in the previous paragraph.

As an exercise in using these definitions we leave the reader to prove the following lemma. (When we return to topology, this lemma will relieve some anxieties about choice of base points.)

(10.4) *If* $g \in G$ *and if* $f: G \to G$ *is a group homomorphism such that* $f(x) = gxg^{-1}$ *for all* x *then* $f_*: Wh(G) \to Wh(G)$ *is the identity map.* □

The groups $Wh(G)$ will be discussed further in the next section.

In computing torsion the following lemma will be quite useful.

(10.5) *If* A, B *and* X *are* $n \times n$, $m \times m$, *and* $n \times m$ *matrices respectively and if* $\tau: GL(R) \to K_G(R)$, *where* G *is any subgroup of the units of* R, *and if* A *has a right inverse or* B *a left inverse, then*

(1) $\begin{pmatrix} A & X \\ 0 & B \end{pmatrix}$ *is non-singular* $\Leftrightarrow$ A *and* B *are non-singular.*

(2) *If* A *and* B *are non-singular then*

$$\tau \begin{pmatrix} A & X \\ 0 & B \end{pmatrix} = \tau(A) + \tau(B).$$

PROOF: (1) holds (for example, when B has a left inverse) because

$$\begin{pmatrix} A & 0 \\ 0 & B \end{pmatrix} = \begin{pmatrix} I_n & -XB^{-1} \\ 0 & I_m \end{pmatrix} \begin{pmatrix} A & X \\ 0 & B \end{pmatrix},$$

where the middle matrix is in $E(R)$ (it is the result of row operations on I_{n+m}), and hence is non-singular.

When A and B are non-singular, the above equation shows that

$$\tau\begin{pmatrix}A & X\\ 0 & B\end{pmatrix} = \tau\begin{pmatrix}A & 0\\ 0 & B\end{pmatrix} = \tau\begin{pmatrix}A & 0\\ 0 & I_m\end{pmatrix} + \tau\begin{pmatrix}I_n & 0\\ 0 & B\end{pmatrix} = \tau(A)+\tau\begin{pmatrix}I_n & 0\\ 0 & B\end{pmatrix}.$$

But a sequence of applications of II_R and then of II_C yields

$$\begin{pmatrix}I_n & 0\\ 0 & B\end{pmatrix} \sim \begin{pmatrix}0 & B\\ (-1)^m I_n & 0\end{pmatrix} \sim \begin{pmatrix}B & 0\\ 0 & (-1)^{2m} I_n\end{pmatrix}$$

Thus $\tau\begin{pmatrix}A & X\\ 0 & B\end{pmatrix} = \tau(A)+\tau(B)$. □

For commutative rings the usual theory of determinants is available and can be used to help keep track of torsion because the elementary operations which take a matrix to another of the same torsion in $K_G(R)$ can only change the determinant by a factor of g for some $g \in G$. A precise statement, the proof of which is left to the reader, is

(10.6) *Suppose that R is a commutative ring and G is a subgroup of the group U of all units of R. Let $SK_1(R) = \tau_G(SL(R))$ where $\tau_G: GL(R) \to K_G(R)$ and $SL(R)$ is the subgroup of $GL(R)$ of matrices of determinant 1. Then there is a split short exact sequence*

$$0 \to SK_1(R) \xrightarrow{\subset} K_G(R) \underset{s}{\overset{[\det]}{\rightleftarrows}} \frac{U}{G} \to 0$$

where $[\det](\tau A) \equiv$ *(the coset of* $(\det A)$ *in* U/G*), and* $s(u\cdot G)$ *is the torsion of the* 1×1 *matrix* (u). *In particular, if R is a field,* $[\det]$ *is an isomorphism.* □

Exercise: The group $SK_1(R) = \tau_G(SL(R))$ defined in (10.6) is independent of G. For we have

$$\underbrace{GL(R) \xrightarrow[\tau_1]{} K_1(R) \xrightarrow{\pi} K_G(R)}_{\tau_G}, \quad \pi \text{ the natural projection,}$$

and $\pi|\tau_1(SL(R)): \tau_1(SL(R)) \xrightarrow{\cong} \tau_G(SL(R))$.

Finally, we close this section with an example due to WHITEHEAD which shows that two $n\times n$ matrices may be equivalent by elementary operations while the equivalence cannot be carried out within the realm of $n\times n$ matrices. Let G be the (non-commutative) group generated by the elements x and y subject to the sole relation that $y^2 = 1$. In $\mathbb{Z}(G)$ let $a = 1-y$ and $b = x(1+y)$. Notice that $ab \neq 0$ while $ba = 0$. Then the 1×1 matrix $(1-ab)$ is not an elementary matrix, but since it represents the same element in $Wh(G)$ as

$$\begin{pmatrix}1-ab & 0\\ 0 & 1\end{pmatrix} = \begin{pmatrix}1 & 0\\ b & 1\end{pmatrix}\begin{pmatrix}1 & a\\ 0 & 1\end{pmatrix}\begin{pmatrix}1 & 0\\ b & 1\end{pmatrix}^{-1}\begin{pmatrix}1 & a\\ 0 & 1\end{pmatrix}^{-1},$$

its torsion is 0.

§11. Some information about Whitehead groups

In the last section we defined, for any group G, the abelian group $Wh(G)$ given by

$$Wh(G) = K_T(\mathbb{Z}G)$$

where T is the group of trivial units of $\mathbb{Z}G$. The computation of WHITEHEAD groups is a difficult and interesting task for which there has developed, in recent years, a rich literature. (See [MILNOR 1], [BASS 1].) We shall content ourselves with first deriving some facts about WHITEHEAD groups of abelian groups which are accessible by totally elementary means and then quoting some important general facts.

(11.1) $Wh(\{1\}) = 0$.

PROOF: This was proven in proving (8.5). □

(11.2) $Wh(\mathbb{Z}) = 0$ [HIGMAN].

PROOF: We think of the group $\mathbb{Z}$ as $\{t^i | i=0, \pm 1, \pm 2, \ldots\}$ so that $\mathbb{Z}(\mathbb{Z})$ is the set of all finite sums $\sum n_i t^i$. Notice that $\mathbb{Z}(\mathbb{Z})$ has only trivial units because the equation

$$(at^\alpha + \ldots + bt^\beta)(ct^\gamma + \ldots dt^\delta) = 1, \ (\alpha \leq \ldots \leq \beta, \gamma \leq \ldots \leq \delta, abcd \neq 0)$$

implies that $\alpha+\gamma = \beta+\delta = 0$. Hence $\alpha = \beta$, $\gamma = \delta$, and these units are trivial.

Suppose that $(a_{ij}(t))$ is an $(n \times n)$-matrix, representing an arbitrary element W of $Wh(\mathbb{Z})$. Multiplying each row by a suitably high power of t, if necessary, we may assume that each entry $a_{ij}(t)$ contains no negative powers of t. Let q be the highest power of t which occurs in any $a_{ij}(t)$. If $q > 1$ then we could obtain another matrix representing W in which the highest power of t which occurs would be $q-1$. For, writing $a_{ij}(t) = b_{ij}(t) + k_{ij}t^q (k_{ij} \in \mathbb{Z})$ we would have

$$(a_{ij}(t)) \sim \begin{pmatrix} (a_{ij}(t)) & t \cdot I_n \\ 0 & I_n \end{pmatrix} \sim \begin{pmatrix} (b_{ij}(t)) & t \cdot I_n \\ (-k_{ij}t^{q-1}) & I_n \end{pmatrix}$$

Thus, proceeding by induction down on q we may assume that, for all i, j, $a_{ij}(t) = b_{ij} + c_{ij}t$ $(b_{ij}, c_{ij} \in \mathbb{Z})$. Thus W may be represented by a matrix—say $m \times m$—with linear entries.

Since $(a_{ij}(t))$ is non-singular its determinant is a unit which, by the first paragraph, is $\pm t^p$ for some p. Expanding the determinant, it follows that either $\det(b_{ij}) = 0$ or $\det(c_{ij}) = 0$. We assume that $\det(b_{ij}) = 0$. (The treatment of the other case is similar.) As is well known, integral row and column operations on the matrix (b_{ij}) will transform it to a diagonal matrix of the form

$$\begin{pmatrix} b'_{11} & & & & & & & \\ & b'_{22} & & & & & & \\ & & \cdot & & & & & \\ & & & \cdot & & & & \\ & & & & b'_{kk} & & & \\ & & & & & 0 & & \\ & & & & & & \cdot & \\ & & & & & & & 0 \end{pmatrix}, \quad k < m$$

Performing the same operations on the matrix $a_{ij}(t)$ leads to $a'_{ij}(t) = b'_{ij} + c'_{ij}t$ where $b'_{ij} = 0$ unless $i = j \leq k$. In particular, the bottom row is of the form

$$(c'_{m1}t \;\; c'_{m2}t \; \ldots \; c'_{mm}t)$$

Multiplying the bottom row by t^{-1} and applying integral column operations (i.e. multiply a column by -1 or add an integral multiple of one column to another), the matrix $(a'_{ij}(t))$ can be transformed into an $n \times n$ matrix $(a''_{ij}(t))$ with linear entries and bottom row of the form

$$(0\ 0, \ldots 0, c''_{mm})$$

for some $c''_{mm} \in \mathbb{Z}$. But $\det(a''_{ij}(t)) = \pm t^p$, so $c''_{mm} = \pm 1$. Hence the last column may be transformed to $\begin{pmatrix} 0 \\ 0 \\ \cdot \\ \cdot \\ 0 \\ 1 \end{pmatrix}$ and W is represented by an $(m-1) \times (m-1)$ matrix with linear entries. Proceeding inductively, W may be represented by a 1×1 matrix (a). But then a is a trivial unit, so $W = 0$. □

More generally, we cite the theorem of [Bass-Heller-Swan]

(11.3) $Wh(\mathbb{Z} \oplus \ldots \oplus \mathbb{Z}) = 0$. □

This is a difficult theorem which has been of great use in recent work ([Kirby-Siebenmann]) in topology in which the n-torus $S^1 \times S^1 \times \ldots \times S^1$ has played a role. (The point is that $\pi_1(S^1 \times \ldots \times S^1) = \mathbb{Z} \oplus \ldots \oplus \mathbb{Z}$, so the Bass-Heller-Swan theorem implies, along with the s-cobordism theorem,[8] that an h-cobordism[8] with an n-torus at one end ($n \geq 5$) is a product). It would be very nice to have a simple geometric proof that $Wh(S^1 \times \ldots \times S^1) = 0$. This is a fact which, as we shall see in §21, is equivalent to (11.3).

We can also exhibit many *non-zero* Whitehead groups. If G is an abelian group, let U be the group of all units of $\mathbb{Z}(G)$, and T the subgroup of trivial units. Then, by (10.6),

$$Wh(G) \cong SK_1(\mathbb{Z}G) \oplus \frac{U}{T}$$

[8] Introduced in §25.

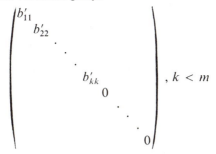

In (8.6) we showed that $Wh(\mathbb{Z}_5) \neq 0$ because $U/T \neq 0$. More generally, we have

(11.4) *If G is an abelian group which contains an element x of order $q \neq 1, 2, 3, 4, 6$ then $Wh(G) \neq 0$. In fact let j, k and a be integers such that*

$$j > 1, k > 1$$

$$j+k < q$$

$$jk = aq \pm 1$$

Such integers always exist. Then a non-trivial unit u of $\mathbb{Z}(G)$ is given by the formula[9]

$$u = (1+x+\ldots+x^{j-1})(1+x+\ldots+x^{k-1})-a(1+x+\ldots+x^{q-1})$$

PROOF: Let $j > 1$ be a prime number less than $(q/2)$ such that j does not divide q. [It is an exercise that such a j exists provided $q \neq 1, 2, 3, 4, 6$.] Then $(j, q) = 1$, so there is an integer $\hat{k}$, $0 < \hat{k} < q$, such that $j\hat{k} = 1 \pmod{q}$. Set $k = \hat{k}$ if $\hat{k} \leq q/2$ and set $k = (q-\hat{k})$ if $\hat{k} > q/2$. Then $jk = aq \pm 1$ for some integer a.

If the element u, given by the formula in the statement of the theorem, is a unit then it certainly is a non-trivial unit. For, since $0 < (j+k-2) < q-1$, either $a = 0$ and $u = 1+(\ldots)+x^{j+k-2}$, or $a \neq 0$ and $u = 1+(\ldots)-ax^{q-1}$.

Consider the case where $jk = aq-1$. Set $\bar{k} = q-k$, $\bar{j} = q-j$ and $\bar{a} = q-k-j+a$. We claim that $uv = 1$ where

$$v = (1+x^j+x^{2j}+\ldots+x^{(\bar{k}-1)j})(1+x^k+\ldots+x^{(\bar{j}-1)k}) - \bar{a}(1+x+x^2+\ldots+x^{q-1})$$

To prove this claim it will suffice to consider the polynomials with integral coefficients, $U(t)$ and $V(t)$, gotten by replacing x by the indeterminate t in the formulas given for u and v respectively, and to show that

$$U(t)V(t) = 1+(t^q-1)P(t)$$

for some polynomial $P(t)$. We shall show in fact that $(t-1)$ and $\Sigma(t) \equiv (1+t+\ldots+t^{q-1})$ both divide $U(t)V(t)-1$.

$U(t)V(t)-1$ is divisible by $(t-1)$ because

$$U(1)V(1)-1 = (jk-aq)(\bar{j}\bar{k}-\bar{a}q)-1 = (-1)(-1)-1 = 0$$

On the other hand

$$U(t) = \left(\frac{t^j-1}{t-1}\right)\left(\frac{t^k-1}{t-1}\right)-a\Sigma(t)$$

$$V(t) = \left(\frac{t^{j\bar{k}}-1}{t^j-1}\right)\left(\frac{t^{k\bar{j}}-1}{t^k-1}\right)-\bar{a}\Sigma(t)$$

[9] This formula is implicit in the general arguments of [Olum 1] and in (12.10) of [Milnor 1].

So

$$U(t)V(t) = \left(\frac{t^{j\bar{k}}-1}{t-1}\right)\left(\frac{t^{k\bar{j}}-1}{t-1}\right) + A(t)\Sigma(t)$$

But each of the quotients shown is of the form $1+B(t)\Sigma(t)$. For $jk \equiv -1 \pmod q$, so we may write $j\bar{k} = 1+bq$ for some b. Then

$$\left(\frac{t^{j\bar{k}}-1}{t-1}\right) = 1+(t+t^2+\ldots+t^q)+(t^{q+1}+\ldots+t^{2q})$$
$$+\ldots+(t^{(b-1)q+1}+\ldots+t^{bq})$$
$$= 1+t\Sigma(t)+t^{q+1}\Sigma(t)+\ldots+t^{(b-1)q+1}\Sigma(t)$$
$$= 1+B(t)\Sigma(t)$$

Thus $U(t)V(t) = [1+B(t)\Sigma(t)][1+C(t)\Sigma(t)]+A(t)\Sigma(t)$ and it follows that $\Sigma(t)$ divides $U(t)V(t)-1$. Thus $uv = 1$.

In the case where $jk = aq+1$, set $\bar{k} = k$, $\bar{j} = j$ and $\bar{a} = a$. The same argument works. □

For cyclic groups (11.4) can be greatly sharpened. In fact we have from [HIGMAN], [BASS 2; p. 54], and [BASS-MILNOR-SERRE; Prop. 4.14],

(11.5) *If $\mathbb{Z}_q$ is the cyclic group of finite order q then*

a) *$Wh(\mathbb{Z}_q)$ is a free abelian group of rank $[q/2]+1-\delta(q)$ where $\delta(q)$ is the number of divisors of q. (In particular $Wh(\mathbb{Z}_q) = 0$ if $q = 1, 2, 3, 4, 6$.)*

b) *$SK_1(\mathbb{Z}(\mathbb{Z}_q)) = 0$ so* (by 10.6) *the determinant map gives an isomorphism* $[\det]: Wh(\mathbb{Z}_q) \to U/T$. □

Added in proof: Let G be a finite group. Then $SK_1(\mathbb{Z}G)$ is finite [BASS 1; p. 625]. But, in contrast to (11.5b), recent work of R. C. Alperin, R. K. Dennis and M. R. Stein shows that, even for finite abelian G, $SK_1(\mathbb{Z}G)$ is usually not zero. For example, $SK_1(\mathbb{Z}_2 \oplus (\mathbb{Z}_3)^3) \cong (\mathbb{Z}_3)^6$.

From (11.5) one sees that the functor $Wh(G)$ does not behave very well with respect to direct products. However for free products we have

(11.6) [STALLINGS 1] *If G_1 and G_2 are any groups then $Wh(G_1 * G_2) = Wh(G_1) \oplus Wh(G_2)$.* □

§12. Complexes with preferred bases [= (R, G)-complexes]

From this point on we assume that G is a subgroup of the units of R which contains the element (-1). For other tacit assumptions the reader is advised to quickly review §9.

An *(R, G)-module* is defined to be a free R-module M along with a "preferred" or "distinguished" family B of bases which satisfies:

If b and b' are bases of M and if $b \in B$ then

$$b' \in B \Leftrightarrow \tau(\langle b/b' \rangle) = 0 \in K_G(R).$$

If M_1 and M_2 are (R, G)-modules and if $f: M_1 \to M_2$ is a module isomorphism then *the torsion of* f—written $\tau(f)$—is defined to be $\tau(A) \in K_G(R)$, where A is the matrix of f with respect to any distinguished bases of M_1 and M_2. One can easily check that $\tau(f)$ is independent of the bases chosen (within the preferred families). We say that f is a *simple isomorphism of* (R, G)-*modules* if $\tau(f) = 0$. In this case we write $f: M_1 \cong M_2(\Sigma)$.

We have introduced the preceding language in order to define an (R, G)-*complex* which is the object of primary interest: An (R, G)-complex is a free chain complex over R

$$C: 0 \to C_n \to C_{n-1} \to \ldots \to C_0 \to 0$$

such that each C_i is an (R, G)-module. A *preferred basis of* C will always mean a basis $c = \bigcup c_i$ where c_i is a preferred basis of C_i.

If G is a group then a $Wh(G)$-*complex* is defined to be an (R, T)-complex, where $R = \mathbb{Z}(G)$ and $T = G \cup (-G)$ is the group of trivial units of $\mathbb{Z}(G)$.

A *simple isomorphism of* (R, G)-*complexes*, $f: C \to C'$, is a chain mapping such that $(f|C_i): C_i \cong C_i'(\Sigma)$, for all i. We write $f: C \cong C'(\Sigma)$. To see that, in fact, this is exactly the right notion of isomorphism in the category of (R, G)-complexes, notice that $f: C \cong C'(\Sigma)$ iff there are preferred bases with respect to which, for each integer i, the matrix of $f|C_i$ is the identity and the matrix of $d_i: C_i \to C_{i-1}$ is identical with that of $d_i': C_i' \to C_{i-1}'$.

Notice that a simple isomorphism of chain complexes is not merely a chain map $f: C \to C'$ which is a simple isomorphism of the (R, G)-modules C and C'. For example, let A be a non-singular $n \times n$ matrix over R with $\tau(A) \neq 0$. Let C_1, C_1', C_2 and C_2' be free modules of rank n with specified preferred bases. Let $C = C_1 \oplus C_2$ and $C' = C_1' \oplus C_2'$. Define $f: C \to C'$ and boundary operators d, d' by the diagram

$$\begin{array}{ccc} C_2 & \xrightarrow{\langle f_2 \rangle = A} & C_2' \\ {\scriptstyle \langle d \rangle = A}\downarrow & & \downarrow{\scriptstyle \langle d' \rangle = A^{-1}} \\ C_1 & \xrightarrow{\langle f_1 \rangle = A^{-1}} & C_1' \end{array}$$

Then f is a chain map and, clearly f is not a simple isomorphism of chain complexes. But f is a module isomorphism with matrix

$$\begin{pmatrix} A & 0 \\ 0 & A^{-1} \end{pmatrix}.$$

Hence by (10.5), $\tau(f) = 0$.

Our purpose in studying chain complexes is to associate to every acyclic (R, G)-complex C a well-defined "torsion element" $\tau(C) \in K_G(R)$ with the following properties:

P1: If $C \cong C'(\Sigma)$ then $\tau(C) = \tau(C')$

P2: If $C' \oplus C''$ is the direct sum of C' and C'' in the category of (R, G)-complexes[10] then $\tau(C' \oplus C'') = \tau(C') + \tau(C'')$

P3: If C is the complex

$$C\colon 0 \to C_n \xrightarrow{d} C_{n-1} \to 0$$

then $\tau(C) = (-1)^{n-1}\tau(d)$

We shall show that, in fact, there is a unique function τ satisfying these properties and that these properties generate a wealth of useful information.

§13. Acyclic chain complexes

In this section we develop some necessary background material concerning acyclic complexes.

An R-module M is said to be *stably free* if there exist free R-modules F_1 and F_2 such that $M \oplus F_1 = F_2$. (Remember, "free" always means with finite basis.)

Notice that if M is a stably free R-module and if $j\colon A \to M$ is a surjection then there is a homomorphism (i.e. a *section*) $s\colon M \to A$ such that $js = 1_M$. For suppose that $M \oplus F$ is free. Then $j \oplus 1\colon A \oplus F \to M \oplus F$ is a surjection and there is certainly a section $S\colon M \oplus F \to A \oplus F$ (gotten by mapping each basis element to an arbitrary element of its inverse image). Then $s = p_1 S i_1$ is the desired section, where $i_1\colon M \to M \oplus F$ and $\pi_1\colon A \oplus F \to A$ are the natural maps.

(13.1) *Suppose that C is a free acyclic chain complex over R with boundary operator d. Denote $B_i = dC_{i+1}$, for all i. Then*

(A) *B_i is stably free for all i.*

(B) *There is a degree-one module homomorphism $\delta\colon C \to C$ such that $\delta d + d\delta = 1$. [Such a homomorphism is called a chain contraction.]*

(C) *If $\delta\colon C \to C$ is any chain contraction then, for each i, $d\delta | B_{i-1} = 1$ and $C_i = B_i \oplus \delta B_{i-1}$.*

REMARK: The δ constructed in proving (B) also satisfies $\delta^2 = 0$, so that there is a pleasant symmetry between d and δ. Moreover, given any chain contraction δ, a chain contraction δ' with $(\delta')^2 = 0$ can be constructed by setting $\delta' = \delta d \delta$.

PROOF: $B_0 = C_0$ because C is acyclic. So B_0 is free. Assume inductively that B_{i-1} is known to be stably free. Then there is a section $s\colon B_{i-1} \to C_i$. Because C is acyclic the sequence

$$0 \to B_i \xrightarrow{\subset} C_i \underset{s}{\overset{d}{\rightleftarrows}} B_{i-1} \to 0$$

is thus a split exact sequence. Hence $C_i = B_i \oplus s(B_{i-1})$ where $s(B_{i-1})$,

[10] i.e., the preferred bases for $(C' \oplus C'')_i$ are determined by a basis which is the union of preferred bases of C'_i and of C''_i.

being isomorphic to B_{i-1}, is stably free. So there exist free modules F_1, F_2 such that $s(B_{i-1}) \oplus F_1 = F_2$. Therefore

$$\begin{aligned} B_i \oplus F_2 &= B_i \oplus s(B_{i-1}) \oplus F_1 \\ &= C_i \oplus F_1. \end{aligned}$$

Since C_i is free, this shows that B_i is stably free, and (A) is proven.

By (A), we may choose, for each surjection $d_i : C_i \to B_{i-1}$, a section $\delta_i : B_{i-1} \to C_i$. As in the proof of (A) it follows that $C_i = B_i \oplus \delta_i(B_{i-1})$. Define $\delta : C \to C$ by the condition that, for all i,

$$\delta | B_i = \delta_{i+1}; \quad \delta | \delta_i(B_{i-1}) = 0.$$

This yields:

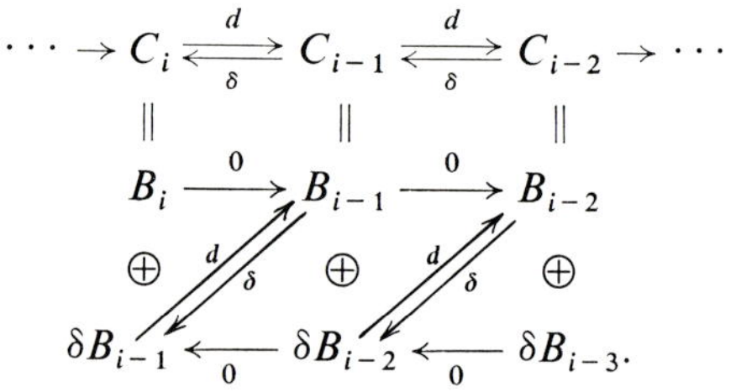

Clearly $d\delta + \delta d = 1$. This proves (B).

Suppose finally we are given $\delta : C \to C$ such that $d\delta + \delta d = 1$. Then $d\delta | B_{i-1} = (d\delta + \delta d) | B_{i-1} = 1_{B_{i-1}}$. Hence $(\delta | B_{i-1}) : B_{i-1} \to C_i$ is a section and, as in (A), $C_i = B_1 \oplus \delta B_{i-1}$. □

(13.2) *If* $0 \to C' \xrightarrow{i} C \xrightarrow{j} C'' \to 0$ *is an exact sequence of chain complexes over* R, *where* C'' *is free and acyclic, then there exists a section* $s : C'' \to C$ *such that* s *is a chain map and* $i + s : C' \oplus C'' \to C$ *is a chain isomorphism.*

PROOF: Let d, d' and d'' be the boundary operators in C, C' and C'' respectively. Let $\delta'' : C'' \to C''$ be a chain contraction. Since each C''_k is free, there are sections $\sigma_k : C''_k \to C_k$. These combine to give a section $\sigma : C'' \to C$.

[Motivation: The map $(\sigma d'' - d\sigma)$ is a homomorphism of degree (-1) which measures the amount by which σ fails to be a chain map. If we wish to add a correction factor to σ—i.e., a degree-zero module homomorphism f such that $\sigma + f$ is a chain map—then a reasonable candidate is the map $f = (d\sigma - \sigma d'')\delta''$. Noticing that then $\sigma + f = d\sigma\delta'' + \sigma\delta'' d''$, we are led to the ensuing argument.]

Let $s = d\sigma\delta'' + \sigma\delta'' d''$. Then $ds = d\sigma\delta'' d'' = sd''$. So s is a chain map. Also

$$\begin{aligned} js = j(d\sigma\delta'' + \sigma\delta'' d'') = jd\sigma\delta'' + \delta'' d'' &= d''(j\sigma)\delta'' + \delta'' d'' \\ &= d''\delta'' + \delta'' d'' = 1. \end{aligned}$$

Thus s is a section. Finally, the isomorphism $i + s$ which comes from the split exact sequence

$$0 \to C' \xrightarrow{i} C \underset{\overset{\curvearrowleft}{s}}{\xrightarrow{j}} C'' \to 0$$

is clearly a chain map, since i and s are chain maps. □

We close this section with a lemma which essentially explains why, in matters concerning torsion, it will not matter which chain contraction of an acyclic complex is chosen.

(13.3) *Suppose that C is an acyclic (R,G)-complex with chain contractions δ and $\bar{\delta}$. For fixed i, let $1 \oplus \delta d$: $C_i \to C_i$ be defined by*

$$1 \oplus \delta d\colon B_i \oplus \bar{\delta} B_{i-1} = C_i \to B_i \oplus \delta B_{i-1} = C_i.$$

Then (A) *$1 \oplus \delta d$ is a simple isomorphism.*

(B) *If B_i and B_{i-1} happen to be free modules with bases b_i and b_{i-1} and if c_i is a basis of C_i then $\tau\langle b_i \cup \delta b_{i-1}/c_i\rangle = \tau\langle b_i \cup \bar{\delta} b_{i-1}/c_i\rangle$.*

PROOF: Denote $g = 1 \oplus \delta d$. Clearly g is an isomorphism since, by (13.1C), $\delta d\colon \bar{\delta} B_{i-1} \xrightarrow{\cong} \delta B_{i-1}$. If B_i and $\bar{\delta} B_{i-1}$ were free then we could, by using a basis for C_i which is the union of a basis for B_i and a basis of $\bar{\delta} B_{i-1}$, write down a matrix which clearly reflects the structure of g. This observation motivates the following proof of (A).

B_i and $\bar{\delta} B_{i-1}$ are stably free, so there exist free modules F_1 and F_2 such that $F_1 \oplus B_i$ and $\bar{\delta} B_{i-1} \oplus F_2$ are free. Fix bases for F_1 and F_2 and take the union of these with a preferred basis for C_i to get a basis c of $F_1 \oplus C_i \oplus F_2$. Let $G = 1_{F_1} \oplus g \oplus 1_{F_2}\colon F_1 \oplus C_i \oplus F_2 \to F_1 \oplus C_i \oplus F_2$. Then

$$\langle G\rangle_{c,c} = \begin{pmatrix} I & & \bigcirc \\ & \langle g\rangle & \\ \bigcirc & & I \end{pmatrix}.$$

So $\tau(\langle G\rangle_{c,c}) = \tau(g)$.

Now choose bases b_1 and b_2 for $(F_1 \oplus B_i)$ and $(\bar{\delta} B_{i-1} \oplus F_2)$ and let $b = b_1 \cup b_2$, another basis for C_i. Notice that $\langle G\rangle_{c,c} = \langle c/b\rangle\langle G\rangle_{b,b}\langle c/b\rangle^{-1}$, so $\tau(\langle G\rangle_{c,c}) = \tau(\langle G\rangle_{b,b})$. But in fact

$$\langle G\rangle_{b,b} = \begin{array}{c} \\ F_1 \oplus B_i \\ \bar{\delta} B_{i-1} \oplus F_2 \end{array} \begin{array}{c|c} F_1 \oplus B_i & \bar{\delta} B_{i-1} \oplus F_2 \\ \hline I & O \\ \hdashline X & I \end{array}.$$

To see this, suppose $y = \bar{\delta} z + w$ where $z \in B_{i-1}$ and $w \in F_2$. Then $G(y) = \delta d(\bar{\delta} z) + w = \delta z + w$. Because $d(\delta z - \bar{\delta} z) = z - z = 0$, we can write $\delta z = \bar{\delta} z + x$ for some $x \in B_i$. Thus $G(y) = (\bar{\delta} z + x) + w = x + y$ where $x \in F_1 \oplus B_i$.

Therefore $\tau(g) = \tau(\langle G\rangle_{c,c}) = \tau(\langle G\rangle_{b,b}) = 0$.

To prove (B), suppose that $b_i = \{u_1, \ldots, u_p\}$ and $b_{i-1} = \{v_1, \ldots v_q\}$ are bases of B_i and B_{i-1} respectively. Let $b = b_i \cup \bar{\delta} b_{i-1}$. Then

$(1 \oplus \delta d)(u_j) = u_j$ and $(1 \oplus \delta d)(\bar{\delta} v_k) = \delta v_k$, and it follows, as pointed out in §9, (page 37), that

$$\langle b_i \cup \delta b_{i-1} / b_i \cup \bar{\delta} b_{i-1} \rangle = \langle 1 \oplus \delta d \rangle_{b,b}.$$

By the proof of (A), $\tau(\langle 1 \oplus \delta d \rangle_{b,b}) = 0$. Thus

$$\begin{aligned} 0 = \tau\langle b_i \cup \delta b_{i-1} / b_i \cup \bar{\delta} b_{i-1} \rangle &= \tau(\langle b_i \cup \delta b_{i-1}/c_i \rangle \langle b_i \cup \bar{\delta} b_{i-1}/c_i \rangle^{-1}) \\ &= \tau\langle b_i \cup \delta b_{i-1}/c_i \rangle - \tau\langle b_i \cup \bar{\delta} b_{i-1}/c_i \rangle. \quad \square \end{aligned}$$

§14. Stable equivalence of acyclic chain complexes

An (R, G)-complex C is defined to an *elementary trivial* complex if it is of the form

$$C\colon 0 \to C_n \xrightarrow{d} C_{n-1} \to 0$$

where d is a simple isomorphism of (R, G)-modules. (Thus, with respect to appropriate preferred bases of C_n and C_{n-1}, $\langle d \rangle$ is the identity matrix.) An (R, G)-complex is *trivial* if it is the direct sum, in the category of (R, G)-complexes, of elementary trivial complexes.

Two (R, G)-complexes C and C' are *stably equivalent*—written $C \overset{s}{\sim} C'$—if there are trivial complexes T and T' such that $C \oplus T \cong C' \oplus T'(\Sigma)$.[11] It is easily checked that this is an equivalence relation.

Just as we showed (§7) that any homotopically trivial CW pair (K, L) is simple-homotopy equivalent to a pair which has cells in only two dimensions, we wish to show that any acyclic (R, G)-complex is stably equivalent to a complex which is zero except in two dimensions.

(14.1) *If C is an acyclic (R, G)-complex of the form*

$$C\colon 0 \to C_n \xrightarrow{d_n} \cdots \xrightarrow{d_{i+3}} C_{i+2} \xrightarrow{d_{i+2}} C_{i+1} \xrightarrow{d_{i+1}} C_i \to 0 \quad (n \geq i+1)$$

and if $\delta\colon C \to C$ is a chain contraction then $C \overset{s}{\sim} C_\delta$ where C_δ is the complex

$$\begin{array}{ccccccccc} C_\delta\colon 0 \to C_n & \xrightarrow{d_n} & \cdots & \xrightarrow{d_{i+3}} & C_{i+2} & \xrightarrow{d_{i+2}} & C_{i+1} & \to 0 \\ & & & & \oplus & \nearrow{\scriptstyle \delta_{i+1}} & & \\ & & & & C_i & & & \end{array}$$

PROOF: For notational simplicity we shall assume that $i = 0$. This will in no way affect the proof.

Let T be the trivial complex with $T_1 = T_2 = C_0$, $T_i = 0$ otherwise, and $\partial_2 = 1\colon T_2 \to T_1$. Let T' be the trivial complex with $T'_0 = T'_1 = C_0$, $T'_i = 0$

[11] WHITEHEAD called this relation "simple equivalence" and wrote $C \equiv C'(\Sigma)$. However, as it is too easily confused with "simple isomorphism" we have adopted the terminology indicated.

otherwise, and $\partial_1' = 1: T_1' \to T_0'$. We claim that $C \oplus T \cong C_\delta \oplus T'(\Sigma)$. The relevant diagrams are

$$
\begin{array}{lcccl}
C \oplus T: \ldots \to C_3 \to & C_2 & & C_1 & \overset{d_1}{\searrow} \\
 & \oplus & \xrightarrow{d_2 \oplus 1} & \oplus & \qquad C_0 \to 0 \\
 & C_0 & & C_0 & \underset{0}{\nearrow}
\end{array}
$$

$$
\begin{array}{lcccccl}
C_\delta \oplus T': \ldots \to C_3 \to & C_2 & \xrightarrow{d_2} & C_1 & \xrightarrow{0} & C_0 \to 0 \\
 & \oplus & \overset{\delta_1}{\nearrow} & \oplus & \overset{1 = \partial_1'}{\nearrow} & \\
 & C_0 & & C_0 & &
\end{array}
$$

Define $f: C \oplus T \to C_\delta \oplus T'$ by

$$f_i = 1, \text{ if } i \neq 1$$

$$f_1(c_0 + c_1) = \delta_1 c_0 + (c_1 + d_1 c_1), \text{ if } c_0 \in C_0 \text{ and } c_1 \in C_1.$$

We leave it to the reader to check that f is a chain map. To show that f is a simple isomorphism we must show that each f_i is a simple isomorphism. This is obvious except for $i = 1$. But

$$
\langle f_1 \rangle = \begin{array}{c} \\ C_0 \\ C_1 \end{array} \!\!\begin{array}{c} \begin{array}{cc} C_0 & C_1 \end{array} \\ \begin{pmatrix} 0 & \langle \delta_1 \rangle \\ \langle d_1 \rangle & I \end{pmatrix} \end{array} = \begin{pmatrix} -I & \langle \delta_1 \rangle \\ 0 & I \end{pmatrix} \begin{pmatrix} I & 0 \\ \langle d_1 \rangle & I \end{pmatrix}.
$$

(This is because, by the conventions of §9, $\langle \delta_1 \rangle \langle d_1 \rangle = \langle d_1 \delta_1 \rangle = \langle d_1 \delta_1 + \delta_0 d_0 \rangle = I$.) Thus, by (10.5), $\tau(f_1) = 0$. □

If C is an acyclic (R, G)-complex an inductive use of (14.1) immediately yields the result that $C \overset{s}{\sim} C'$ for some (R, G)-complex C' which is 0 except in two dimensions. This is the only consequence of (14.1) which we will use. However, in order to motivate the definition to be given in the next section we give here a precise picture (at least in the case when $\delta^2 = 0$) of the complex C' which is constructed by repeated application of (14.1).

(14.2) *Let C be an acyclic (R, G)-complex with boundary operator d and chain contraction δ satisfying $\delta^2 = 0$ and let*

$$C_{\text{odd}} = C_1 \oplus C_3 \oplus \ldots$$

$$C_{\text{even}} = C_0 \oplus C_2 \oplus \ldots .$$

Then C is stably equivalent to an (R, G)-complex of the form

$$C': 0 \to C_m' = C_{\text{odd}} \xrightarrow{(d+\delta)|C_{\text{odd}}} C_{m-1}' = C_{\text{even}} \to 0$$

for some odd integer m.

PROOF: Let m be an odd integer such that C is of the form $0 \to C_m \to C_{m-1} \to \ldots \to C_0 \to 0$. (We allow the possibility that $C_m = 0$.) If $j \leq m$, let

$C'_j = C_j \oplus C_{j-2} \oplus C_{j-4} \oplus \ldots$. Let D^i be the chain complex

$$D^i : 0 \to C_m \xrightarrow{d} \ldots \to C_{i+2} \xrightarrow{d} C'_{i+1} \xrightarrow{d'} C'_i \to 0$$

where d is the boundary operator in C and $d' = (d|C_{i+1}) + (d+\delta)|C'_{i-1}$. It is easily checked that D^i is a chain complex, $D^0 = C$ and $D^{m-1} = C'$. Now define $\Delta^i = \Delta : D^i \to D^i$, a degree-one homomorphism, by

$$\Delta|C_j = \delta|C_j \quad \text{if} \quad j \geq i+2,$$

$\Delta|C'_{i+1} = (\delta|C_{i+1}) \circ \pi$, where $\pi : C'_{i+1} \to C_{i+1}$ is the natural projection,

$$\Delta|C'_i = (d+\delta)|C'_i.$$

Then Δ is a chain contraction of D^i. This is easily checked once one notes (with d^i denoting the boundary operator of D^i) that

$$\begin{aligned}(d^i\Delta + \Delta d^i)|C'_{i+1} &= (d\delta + \delta d)|C_{i+1} \oplus (d+\delta)^2|C'_{i-1} \\ &= 1_{C'_{i+1}}\end{aligned}$$

since $d^2 = \delta^2 = 0$.

Thus D^i and Δ^i satisfy the hypothesis of (14.1). The conclusion of (14.1) says precisely that $D^i \overset{s}{\sim} D^{i+1}$. By induction, $C \overset{s}{\sim} C'$. □

§15. Definition of the torsion of an acyclic complex

Motivated by (14.2) we make the following definition:

Let C be an acyclic (R, G)-complex with boundary operator d. Let δ be any chain contraction of C. Set

$$C_{\text{odd}} = C_1 \oplus C_3 \oplus \ldots$$

$$C_{\text{even}} = C_0 \oplus C_2 \oplus \ldots$$

$$(d+\delta)_{\text{odd}} = (d+\delta)|C_{\text{odd}} : C_{\text{odd}} \to C_{\text{even}}.$$

Then $\tau(C) \equiv \tau((d+\delta)_{\text{odd}}) \in K_G(R)$.

In particular, if C is a $Wh(G)$-complex then $\tau(C) = \tau((d+\delta)_{\text{odd}}) \in Wh(G)$.

Unlike (14.2), the definition does not assume that $\delta^2 = 0$. This would be a totally unnecessary assumption, although it would be a modest convenience in proving that $\tau(C)$ is well-defined.

We shall write $d+\delta$ instead of $(d+\delta)_{\text{odd}}$ when no confusion can occur. It is understood that $\tau(C)$ is defined in terms of preferred bases of C_{odd} and C_{even}. If $c = \bigcup c_i$ is a preferred basis of C then $\tau(C) = \tau(\langle d+\delta \rangle_{c_{\text{odd}}, c_{\text{even}}})$ where $c_{\text{odd}} = (c_1 \cup c_3 \cup \ldots)$ and $c_{\text{even}} = (c_0 \cup c_2 \cup \ldots)$. For convenience $\langle d+\delta \rangle_{c_{\text{odd}}, c_{\text{even}}}$ will be abbreviated to $\langle d+\delta \rangle_c$ or simply to $\langle d+\delta \rangle$ when the context is unambiguous.

The form of $\langle d+\delta\rangle$ is

$$\begin{array}{c} \\ C_1 \\ C_3 \\ C_5 \end{array}\begin{array}{c} \begin{array}{ccc} C_0 & C_2 & C_4 \end{array} \\ \hline \left(\begin{array}{cccc} \langle d_1\rangle & \langle\delta_2\rangle & & 0 \\ & \langle d_3\rangle & \langle\delta_4\rangle & \\ 0 & & \langle d_5\rangle & \\ & & & \ddots \end{array}\right) \end{array}$$

However, the reader must not fall into the trap of concluding that "$\tau(d+\delta) = \tau(d_1)+\tau(d_3)+\ldots$". For the matrices $\langle d_i\rangle$ are usually not invertible, or even square.

To show that $\tau(C)$ is well-defined, we must show that $\langle d+\delta\rangle$ is non-singular and that $\tau(d+\delta)$ is independent of which preferred basis is used and of which chain contraction δ is used.

(15.1) *Let c be a basis for C. Then $\langle(d+\delta)_{\text{odd}}\rangle_c$ and $\langle(d+\delta)_{\text{even}}\rangle_c$ are non-singular with $\tau(\langle(d+\delta)_{\text{odd}}\rangle_c) = -\tau(\langle(d+\delta)_{\text{even}}\rangle_c)$.*

PROOF: $(d+\delta)_{\text{even}} \circ (d+\delta)_{\text{odd}} = (d^2+d\delta+\delta d+\delta^2)|C_{\text{odd}} = (1+\delta^2)|C_{\text{odd}}$. Therefore

$$\langle(d+\delta)_{\text{odd}}\rangle_c\langle(d+\delta)_{\text{even}}\rangle_c = \begin{array}{c} \\ C_1 \\ C_3 \\ C_5 \end{array}\begin{array}{c} \begin{array}{cc} C_1 & C_3 \end{array} \\ \hline \left(\begin{array}{cccc} I & \langle\delta_3\delta_2\rangle & & 0 \\ & I & \langle\delta_5\delta_4\rangle & \\ 0 & & I & \\ & & & \ddots \end{array}\right) \end{array}$$

By (10.5) this matrix is non-singular and has zero torsion. A similar assertion holds for $\langle(d+\delta)_{\text{even}}\rangle_c\langle(d+\delta)_{\text{odd}}\rangle_c$ and the result follows. □

(15.2) *Let $c = \bigcup c_i$ and $c' = \bigcup c_i'$ be bases of $\mathbf{C}$, where the c_i and c_i' are arbitrary bases of C_i. Then*

$$\tau(\langle d+\delta\rangle_c) = \tau(\langle d+\delta\rangle_{c'}) + \sum_i (-1)^i\tau(\langle c_i'/c_i\rangle).$$

In particular, if the c_i and c_i' are preferred bases, $\tau(\langle d+\delta\rangle_c) = \tau(\langle d+\delta\rangle_{c'})$, so $\tau(d+\delta)$ is independent of which preferred basis is used.

PROOF: $\langle d+\delta\rangle_c = \langle c_{\text{odd}}/c'_{\text{odd}}\rangle\langle d+\delta\rangle_{c'}\langle c'_{\text{even}}/c_{\text{even}}\rangle$

$$= \left(\begin{array}{ccc} \langle c_1/c_1'\rangle & & 0 \\ & \langle c_3/c_3'\rangle & \\ 0 & & \ddots \end{array}\right)\langle d+\delta\rangle_{c'}\left(\begin{array}{ccc} \langle c_0'/c_0\rangle & & 0 \\ & \langle c_2'/c_2\rangle & \\ 0 & & \ddots \end{array}\right)$$

Therefore, using (10.5),

$$\begin{aligned}\tau(\langle d+\delta\rangle_c) &= \sum_i \tau\langle c_{2i+1}/c'_{2i+1}\rangle + \tau(\langle d+\delta\rangle_{c'}) + \sum_i \tau\langle c'_{2i}/c_{2i}\rangle \\ &= \sum_i \tau(\langle c'_{2i+1}/c_{2i+1}\rangle^{-1}) + \tau(\langle d+\delta\rangle_{c'}) + \sum_i \tau\langle c'_{2i}/c_{2i}\rangle \\ &= \tau(\langle d+\delta\rangle_{c'}) + \sum_i (-1)^i \tau(c'_i/c_i\rangle. \quad \square\end{aligned}$$

(15.3) *Suppose that C is an acyclic (R, G)-complex with chain contractions δ and $\bar{\delta}$. Then $\tau(d+\delta) = \tau(d+\bar{\delta})$.*

PROOF: (All calculations will be with respect to a fixed preferred basis.)

$$\begin{aligned}\tau(d+\bar{\delta}) - \tau(d+\delta) &= \tau\langle(d+\bar{\delta})_{\text{odd}}\rangle + \tau\langle(d+\delta)_{\text{even}}\rangle, \text{ by (15.1)} \\ &= \tau\langle(d+\delta)\circ(d+\bar{\delta})|C_{\text{odd}}\rangle \\ &= \tau\langle(\delta d + d\bar{\delta} + \delta\bar{\delta})|C_{\text{odd}}\rangle\end{aligned}$$

$$= \text{torsion of } \begin{array}{c|ccc} & C_1 & C_3 & C_5 \\ \hline C_1 & \langle\delta d + d\bar{\delta}\rangle & \langle\delta\bar{\delta}\rangle & 0 \\ C_3 & 0 & \langle\delta d + d\bar{\delta}\rangle & \langle\delta\bar{\delta}\rangle \\ C_5 & & & \langle\delta d + d\bar{\delta}\rangle \; . \; . \quad \langle\delta\bar{\delta}\rangle \; . \; . \end{array}$$

$= \sum_i \tau\langle(\delta d + d\bar{\delta})|C_{2i+1}\rangle$, by (10.5), if each $(\delta d + d\bar{\delta})|C_{2i+1}$ is non-singular.

Notice, however, that $(\delta d + d\bar{\delta})|C_j = 1 \oplus \delta d: B_j \oplus \bar{\delta}B_{j-1} \to B_j \oplus \delta B_{j-1}$. (Here $B_k = dC_{k+1}$.) For, if $b_j \in B_j$ and $b_{j-1} \in B_{j-1}$ we have

$$\begin{aligned}(\delta d + d\bar{\delta})(b_j) &= b_j \\ (\delta d + d\bar{\delta})(\bar{\delta}b_{j-1}) &= (\delta d)(\bar{\delta}b_{j-1}) + (1 - \bar{\delta}d)(\bar{\delta}b_{j-1}) \\ &= \delta d(\bar{\delta}b_{j-1}) + \bar{\delta}b_{j-1} - \bar{\delta}b_{j-1} \\ &= \delta d(\bar{\delta}b_{j-1}).\end{aligned}$$

Hence, by (13.3A), $(\delta d + d\bar{\delta})|C_{2i+1}$ is a simple isomorphism. Thus $\tau(d+\bar{\delta}) = \tau(d+\delta)$. $\square$

This completes the proof that $\tau(C)$ is well-defined.

§16. Milnor's definition of torsion

In [Milnor 1] the torsion of an acyclic (R, G)-complex C with boundary operator d is formulated as follows:

For each integer i, let $B_i = dC_{i+1}$ and let c_i be a preferred basis for C_i. Let F_i be a free module with a distinguished basis such that $B_i \oplus F_i$ is also free. For notational convenience, set $G_i = F_{i-1}$. Choose bases b_i for $B_i \oplus F_i$

in an arbitrary manner and let c_i' be the natural distinguished basis for $C_i \oplus F_i \oplus G_i$. Then

$$0 \to B_i \oplus F_i \xrightarrow{\subset} C_i \oplus F_i \oplus G_i \xrightarrow{d_i \oplus 0 \oplus 1} B_{i-1} \oplus F_{i-1} \to 0$$

is an exact sequence of free modules. Let $\Delta_i : B_{i-1} \oplus F_{i-1} \to C_i \oplus F_i \oplus G_i$ be a section. Set $b_i b_{i-1} = b_i \cup \Delta_i(b_{i-1})$, a basis for $C_i \oplus F_i \oplus G_i$. MILNOR's torsion, $\tau_M(C)$, is defined by

$$\tau_M(C) = \sum_i (-1)^i \, \tau\langle b_i b_{i-1}/c_i' \rangle.$$

(16.1) *If C is an acyclic (R, G)-complex then $\tau(C) = \tau_M(C)$.*

PROOF: Let T_i be the trivial complex $0 \to G_{i+1} \xrightarrow{1} F_i \to 0$, where T_i is 0 except in dimensions i and $i+1$, and $G_{i+1} = F_i$ with the same distinguished basis. Let $C' = C \oplus T_0 \oplus T_1 \oplus \ldots T_{n-1}$ ($n = \dim C$). We claim that $\tau(C') = \tau(C)$. For let d' be the boundary operator in C'. Let δ be a chain contraction of C and let $\delta' = \delta \oplus \varepsilon_0 \oplus \ldots \oplus \varepsilon_{n-1}$ where $\varepsilon_i : F_i \xrightarrow{1} G_{i+1}$. Clearly δ' is a chain contraction of C'. Moreover, $\langle d' + \delta' \rangle$ is gotten from $\langle d + \delta \rangle$ simply by adding identity blocks of the form

$$\begin{array}{c|c|c|c} & & F_{2i} & \\ \hline & & 0 & \\ \hline G_{2i+1} & 0 & I & 0 \\ \hline & & 0 & \end{array} \qquad \text{and} \qquad \begin{array}{c|c|c|c} & & G_{2i+2} & \\ \hline & & 0 & \\ \hline F_{2i+1} & 0 & I & 0 \\ \hline & & 0 & \end{array}$$

These cannot change the torsion, so $\tau(C) = \tau(C')$. It remains to show that $\tau(C') = \tau_M(C)$.

Let $\Delta_i : B_{i-1} \oplus F_{i-1} \to (C_i \oplus F_i \oplus G_i) = C_i'$ be the section and c_i' the preferred basis of C_i' given in the definition of $\tau_M(C)$. Let $B_i' = d'(C_{i+1}) = B_i \oplus F_i$. So $C_i' = B_i' \oplus \Delta_i B_{i-1}'$. Let the chain contraction $\Delta : C' \to C'$ be given by

$$\Delta | B_i' = \Delta_{i+1}; \quad \Delta | \Delta_i B_{i-1}' = 0.$$

Set $b = \bigcup_i (b_i b_{i-1})$, a basis of C'. Then $\tau(\langle d' + \Delta \rangle_b) = 0$ because the basis b has been chosen so that, for each i,

$$\langle (d' + \Delta) | C_{2i+1} \rangle_b = \begin{array}{c|c|cc|c} & & B_{2i}' & \Delta B_{2i+1}' & \\ \hline B_{2i+1}' & 0 & 0 & I & 0 \\ \Delta B_{2i}' & & I & 0 & \end{array}$$

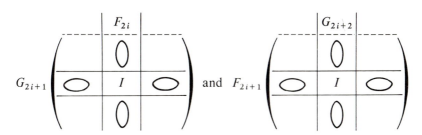

Letting $c' = \bigcup c_i'$, (15.2) then gives

$$\begin{aligned}\tau(C') = \tau(\langle d'+\Delta\rangle_{c'}) &= \tau(\langle d'+\Delta\rangle_b) + \sum(-1)^i\tau\langle b_ib_{i-1}/c_i'\rangle \\ &= \sum(-1)^i\tau\langle b_ib_{i-1}/c_i'\rangle \\ &= \tau_M(C). \quad \square\end{aligned}$$

Though we shall not use the greater generality, it is interesting to note how MILNOR used this formulation to define $\tau(C)$ not only when $H_*(C) = 0$, but also in the case when each $H_i(C)$ is free with a given preferred basis h_i and each C_i is free with preferred basis c_i. In these circumstances one has short exact sequences

$$0 \to Z_i \to C_i \to B_{i-1} \to 0$$

$$0 \to B_i \to Z_i \to H_i \to 0.$$

Arguing as in (13.1) all the B_i and Z_i can be seen to be stably free, and there are sections $\delta_i: B_{i-1} \to C_i$ and $s_i: H_i \to Z_i$. Thus

$$C_i = Z_i \oplus \delta_i B_{i-1} = s_i(H_i) \oplus B_i \oplus \delta_i B_{i-1}.$$

Clearly $s_i(h_i)$ is a basis of $s_i(H_i)$. Choose $F_i = G_{i+1}$ so that $B_i \oplus F_i$ is free. Let $\Delta_i = \delta_i \oplus 1_{F_{i-1}}$, and let b_i be any basis of $B_i \oplus F_i$. Then $s_i(h_i) \cup b_i \cup \Delta_i(b_{i-1})$ is a basis for $C_i \oplus F_i \oplus G_i$. This basis is denoted $b_ih_ib_{i-1}$. Again let c_i' denote the trivial extension of c_i to a basis of $C_i \oplus F_i \oplus G_i$. Then MILNOR defines

$$\tau(C) = \sum_i(-1)^i\tau\langle b_ih_ib_{i-1}/c_i'\rangle.$$

For more details the reader is referred to [MILNOR 1].

§17. Characterization of the torsion of a chain complex

In this section we prove (as promised earlier) that the torsion operator τ satisfies properties **P1–P3** below and is, in fact, the only operator to do so. Moreover τ induces an isomorphism of stable equivalence classes of acyclic (R, G)-complexes with $K_G(R)$.

(17.1) *If R is a ring and G is a subgroup of the units of R containing (-1), and if $\mathscr{C}$ is the class of acyclic (R, G)-complexes, then the torsion map $\tau:\mathscr{C} \to K_G(R)$ defined in §15 satisfies*

P1: $C \cong C'(\Sigma) \Rightarrow \tau(C) = \tau(C')$

P2: $\tau(C' \oplus C'') = \tau(C') + \tau(C'')$

P3: $\tau(0 \to C_n \xrightarrow{d} C_{n-1} \to 0) = (-1)^{n-1}\tau(d)$.

PROOF: In what follows, d, d', d'' denote boundary operators for C, C', C'' respectively.

Suppose that $f: C \cong C'(\Sigma)$. As pointed out in §12, this means that there are distinguished bases of C and C' such that $\langle d_n \rangle = \langle d_n' \rangle$ and $\langle f_n \rangle = I$, for all n. Choose a chain contraction $\delta: C \to C$ and let $\delta' = f\delta f^{-1}$. Then $\langle d+\delta \rangle = \langle d'+\delta' \rangle$. Therefore $\tau(C) = \tau(C')$ and **P1** is verified.

Assume that $C = C' \oplus C''$ (so $d = d' \oplus d''$) and that δ', δ'' are chain contractions for C' and C''. It follows that $\delta = \delta' \oplus \delta''$ is a chain contraction for C. Since permutation of rows or columns does not change the torsion of a matrix, we have

$$\begin{aligned}\tau(C) &= \tau\langle d+\delta \rangle = \tau\langle (d' \oplus d'') + (\delta' \oplus \delta'') \rangle \\ &= \tau \begin{pmatrix} \langle d'+\delta' \rangle & \bigcirc \\ \bigcirc & \langle d''+\delta'' \rangle \end{pmatrix} \\ &= \tau(C') + \tau(C'').\end{aligned}$$

Finally, suppose that C is:

$$0 \to C_n \xrightarrow{d} C_{n-1} \to 0.$$

Set $\delta_j = 0$ if $j \neq n$ and $\delta_n = d_n^{-1}: C_{n-1} \to C_n$. If n is odd, $\delta | C_{\text{odd}} = 0$ so $\tau C = \tau(d) = (-1)^{n-1}\tau(d)$. We leave the case where n is even to the reader. □

The property **P2**, as stated, is too restrictive for practical situations where more general short exact sequences usually occur. We diverge briefly to prove a more general form of **P2** which we call $\overline{\mathbf{P2}}$.

(17.2) *Suppose that* $0 \to C' \xrightarrow{i} C \xrightarrow{j} C'' \to 0$ *is a short exact sequence of acyclic chain complexes and that* $\sigma: C'' \to C$ *is a degree-zero section (but not necessarily a chain map). Assume further that* C, C' *and* C'' *are* (R, G)*-complexes with preferred bases* c, c' *and* c''. *Then*

$$\tau(C) = \tau(C') + \tau(C'') + \sum (-1)^k \tau\langle c_k' c_k'' / c_k \rangle$$

where $c_k' c_k'' \equiv i(c_k') \cup \sigma(c_k'')$. *In particular, if* $i(c_k') \cup \sigma(c_k'')$ *is a preferred basis of* C_k *for all* k, *then* $\tau(C) = \tau(C') + \tau(C'')$.

PROOF: By (13.2) there is a chain map section $s: C'' \to C$ such that $i+s: C' \oplus C'' \to C$ is an isomorphism. The bases $\gamma_k = i(c_k') \cup s(c_k'')$ of the C_k make C into a new (R, G) complex C^γ. Clearly $i+s: C' \oplus C'' \cong C^\gamma(\Sigma)$. Hence, using property **P2**, $\tau(C^\gamma) = \tau(C') + \tau(C'')$. Let δ be a chain contraction of C. Then,

$$\begin{aligned}\tau(C) &= \tau(\langle d+\delta \rangle_c) \\ &= \tau(\langle d+\delta \rangle_\gamma) + \sum_k (-1)^k \tau\langle \gamma_k / c_k \rangle, \quad \text{by (15.2)} \\ &= \tau(C^\gamma) + \sum (-1)^k \tau\langle \gamma_k / c_k \rangle \\ &= \tau(C') + \tau(C'') + \sum (-1)^k \tau\langle \gamma_k / c_k \rangle.\end{aligned}$$

Finally, $\tau\langle c_k' c_k'' / c_k \rangle = \tau\langle \gamma_k / c_k \rangle$. For, the short exact sequence of free

(R, G)-modules $0 \to C_k' \to C_k \to C_k'' \to 0$ may be thought of as an acyclic (R, G)-complex, and the result follows from (13.3B). □

(17.3) *If $\mathscr{C}$ is the class of acyclic (R, G)-complexes then the torsion map $\tau: \mathscr{C} \to K_G(R)$ is the only function satisfying properties* **P1–P3** *of* (17.1).

PROOF: By (17.1), τ does satisfy these properties. Suppose that $\mu: \mathscr{C} \to K_G(R)$ also does so. Using (14.2), $C \overset{s}{\sim} C'$ where C' is $0 \to C_m' \xrightarrow{d'} C_{m-1}' \to 0$. Then $C \oplus T \cong C' \oplus T'(\Sigma)$ for some trivial complexes T, T'. Properties **P1–P3** imply that $\tau(C) = \tau(C)+\tau(T) = \tau(C')+\tau(T') = \tau(C')$. Similarly $\mu(C) = \mu(C')$. But by **P3**, $\tau(C') = (-1)^{m-1}\tau(d') = \mu(C')$. Thus $\tau(C) = \mu(C)$. □

(17.4) *Let $\mathscr{C}_0$ be the set of all stable equivalence classes of acyclic (R, G)-complexes, viewed as a semi-group under the operation*

$$[C] + [C'] = [C \oplus C']$$

where $[C]$ denotes the equivalence class of C. Let $\tau_0: \mathscr{C}_0 \to K_G(R)$ by

$$\tau_0[C] = \tau(C).$$

Then $\mathscr{C}_0$ is a group and τ_0 is a group isomorphism.[12]

PROOF: Since $K_G(R)$ is a group, the result will follow once we show that τ_0 is a semi-group isomorphism.

The proof that τ_0 is a well-defined, surjective homomorphism is left to the reader.

To see that τ_0 is one-one, suppose that $\tau_0[C] = \tau_0[D]$. Choose an odd integer $p > \max\{\dim C, \dim D\}$. Repeated use of (14.1) allows us to assert that

$$C \overset{s}{\sim} C' = (0 \to C_p' \xrightarrow{d'} C_{p-1}' \to 0)$$

$$D \overset{s}{\sim} D' = (0 \to D_p' \xrightarrow{\Delta} D_{p-1}' \to 0).$$

Adding a trivial complex to C' or D' if necessary, we may assume that C_p' and D_p' have equal rank. Choose distinguished bases for C' and D'. We define $f: C' \to D'$ by defining $f_p: C_p' \to D_p'$ to satisfy the condition that $\langle f_p \rangle = I$ and by setting $f_{p-1} = \Delta f_p (d')^{-1}$. Clearly f is a chain isomorphism. Also $\tau(f_p) = 0$ and $\tau(f_{p-1}) = \tau(\Delta)+\tau(f_p)-\tau(d') = \tau_0[D]-\tau_0[C] = 0$. Thus f is a simple isomorphism. This proves that $[C] = [C'] = [D'] = [D]$, so τ_0 is one-one. □

§18. Changing rings

As usual, (R, G) and (R', G') each denotes a ring and a subgroup of the units of this ring which contains -1.

---

[12] This theorem has also been observed by [Cockroft-Combes].

If C is an (R, G)-complex and $h: R \to R'$ is a ring homomorphism with $h(G) \subset G'$ then we may construct an (R', G')-complex C_h as follows: Choose a preferred basis $c = \{c_k^i\}$ for C, and let C_h be the free graded R'-module generated by the set c. We denote $c = \hat{c}$ when c is being thought of as a subset of C_h. Define $\hat{d}: C_h \to C_h$ by setting $\hat{d}(\hat{c}_k^i) = \sum_j h(a_{kj})\hat{c}_j^{i-1}$ if $d(c_k^i) = \sum_j a_{kj}c_j^{i-1}$. We stipulate that $\hat{c}$ is a preferred basis of C_h, thus making $\boldsymbol{C_h}$ into an (R', G')-complex. [That C_h is independent, up to simple isomorphism, of the choice of c follows from the fact that the induced map $h_*: GL(R) \to GL(R')$ takes matrices of 0 torsion to matrices of 0 torsion. This will be made clear in step 6 at the end of this section when we redefine C_h as $R' \otimes_h C$.]

This change of rings is useful for several reasons. One reason is that C_h may be acyclic even when C is not. Thus we gain an algebraic invariant for C—namely $\tau_h(C)$—defined by

$$\tau_h(C) = \tau(C_h) \in K_{G'}(R')$$

The following example of this phenomenon occurs in the study of lens spaces.

(18.1) *Suppose that*

$\mathbb{Z}_p = \{1, t, \ldots, t^{p-1}\}$, *a cyclic group of order* p $(1 < p \in \mathbb{Z})$

$R = \mathbb{Z}(\mathbb{Z}_p)$

$G = \{\pm t^j | j \in \mathbb{Z}\} \subset R$

$R' = \mathbb{C}$ (*the field of complex numbers*)

ξ *is a* p'th *root of unity*; $\xi \neq 1$

$G' = \{\pm \xi^j | j \in \mathbb{Z}\} \subset R'$

$(r_1, \ldots, r_n) =$ *a sequence of integers relatively prime to* p

$\Sigma(t) = 1+t+\ldots+t^{p-1} \in R$

$h: R \to R'$ by $h(\sum_j n_j t^j) = \sum_j n_j \xi^j$

Suppose further that C *is the* (R, G)*-complex*

$$0 \to C_{2n-1} \xrightarrow{\langle t^{r_n}-1\rangle} C_{2n-2} \xrightarrow{\langle \Sigma(t)\rangle} C_{2n-3} \xrightarrow{\langle t^{r_{n-1}}-1\rangle}$$
$$C_{2n-4} \xrightarrow{\langle \Sigma(t)\rangle} \cdots \xrightarrow{\langle t^{r_1}-1\rangle} C_0 \to 0$$

where each $C_j (0 \leq j \leq 2n-1)$ *has rank* 1 *and the* 1×1 *matrix of* d_j *is written above the arrow* $C_j \to C_{j-1}$. *Then* C *is not acyclic while* C_h *is acyclic with* $\tau(C_h) \in K_{G'}(\mathbb{C})$ equal to the torsion of the 1×1 matrix $\langle \prod_{j=1}^n (\xi^{r_j} - 1)\rangle$.

PROOF: C is a chain complex because

$$\Sigma(t)\cdot(t^{r_j}-1) = \Sigma(t)(t-1)(1+t+\ldots+t^{r_j-1}) = (t^p-1)(1+\ldots+t^{r_j-1}) = 0.$$

C is not acyclic. For, if $\{c_i\}$ is a basis for C_i, then $\Sigma(t)\cdot c_{2n-1}$ is not a boundary while $d[\Sigma(t)\cdot c_{2n-1}] = \Sigma(t)\cdot(t^{r_n}-1)c_{2n-2} = 0$, so it is a cycle. However C_h is acyclic. For $(1+\xi+\ldots+\xi^{p-1}) = [(\xi^p-1)/(\xi-1)] = 0$, and consequently C_h is of the form

$$0 \to C_{2n-1} \xrightarrow{\langle \xi^{r_n}-1\rangle} C_{2n-2} \xrightarrow{0} C_{2n-3} \xrightarrow{\langle \xi^{r_{n-1}}-1\rangle} C_{2n-4} \xrightarrow{0} \ldots$$

Since r_j is prime to p, we have $\xi^{r_j} \neq 1$. Thus each matrix $\langle \xi^{r_j}-1\rangle$ is non-singular and C_h is acyclic. It is an exercise for the reader that $\tau(C_h) = \tau\left\langle \prod_{j=1}^{n} (\xi^{r_j}-1)\right\rangle$. □

Sometimes, when C is acyclic, $\tau(C)$ is very hard to compute while $\tau_h(C)$ is very easy to compute. When such a homomorphism h can be found it often pays to change rings because of

(18.2) *If C is an acyclic (R, G)-complex and $h:(R, G) \to (R', G')$ is a ring homomorphism then C_h is acyclic and $\tau_h(C) = h_*\tau(C)$ where $h_*: K_G(R) \to K_{G'}(R')$ is the induced map.*

PROOF: Choose a chain contraction δ of C and suppose that $\delta(c_k^i) = \sum_j b_{kj}c_j^{i+1}$. Define $\hat{\delta}: C_h \to C_h$ by $\hat{\delta}(\hat{c}_k^i) = \sum_j h(b_{kj})\hat{c}_j^{i+1}$. Clearly, since $h(G) \subset G'$, we have $h(1) = 1$ so that: $\langle \hat{d}\hat{\delta}+\hat{\delta}\hat{d}\rangle = h_*\langle d\delta+\delta d\rangle = h_*(I) = I$, and $h_*\langle d+\delta\rangle = \langle \hat{d}+\hat{\delta}\rangle$. Thus $\hat{\delta}$ is a chain contraction and $\tau_h(C) = \tau(C_h) = h_*\tau(C)$. □

As a simple but important application of (18.2) we have

(18.3) *Suppose that C is an acyclic $Wh(G)$-complex with boundary operator d. If there is a preferred basis c of C with respect to which $\langle d\rangle$ has only integral entries.* (i.e. $\langle d\rangle = (a_{ij})$ where $a_{ij} \in \mathbb{Z} \subset \mathbb{Z}(G)$) *then $\tau(C) = 0$.*

PROOF: Let $C' \subset C$ be the free $\mathbb{Z}$-module generated by c. Let $d' = d|C'$. Since d is integral, $d': C' \to C'$, so that C' becomes a chain complex and, indeed, a free $(\mathbb{Z}, \{\pm 1\})$-complex if we specify c as preferred basis. Let $h:(\mathbb{Z}, \{\pm 1\}) \to (\mathbb{Z}(G), G \cup -G)$ be the inclusion map. Clearly we can identify $C \equiv C'_h$.

We claim that C' is acyclic. For suppose that $d'(x) = 0$ where $x \in C'_i$. Then $x = d(y)$ for some $y \in C_{i+1}$, since C is acyclic. Suppose that $x = \sum_k n_k c_k^i$ and $y = \sum_j r_j c_j^{i+1}$ $(n_k \in \mathbb{Z}, r_j \in \mathbb{Z}(G))$. We have

$$d(y) = \sum r_j \sum_k a_{jk}c_k^i \qquad (a_{jk} \in \mathbb{Z})$$

so

$$n_k = \sum r_j a_{jk}$$

Let $A:\mathbb{Z}(G) \to \mathbb{Z}$ be the ring homomorphism given by $A(\sum_j m_j g_j) = \sum_j m_j$ and set $y' = \sum_j A(r_j)c_j^{i+1} \in C'_{i+1}$.

Then

$$\begin{aligned} d'(y') &= \sum_{j,k} A(r_j)a_{jk}c'^{i}_{k} \\ &= \sum_{j,k} A(r_j a_{jk})c'^{i}_{k} \\ &= \sum_{k} A(\sum_{j} r_j a_{jk})c'^{i}_{k} \\ &= \sum_{k} A(n_k)c'^{i}_{k} \\ &= \sum_{k} n_k c'^{i}_{k} = x \end{aligned}$$

Thus every cycle is a boundary and C' is acyclic.

By (18.2), $\tau(C) = \tau(C'_h) = h_*\tau(C')$. But $\tau(C') \in Wh(\{1\}) = 0$. Hence $\tau(C) = 0$. □

To put the construction $C \mapsto C_h$ into proper perspective and to allow ourselves access to well-known algebraic facts in dealing with it, we now outline a richer description of C_h. We leave the reader to check the elementary assertions about tensor algebra being used. (A good reference is [CHEVALLEY; Ch. III, §8, 11])

1. R' becomes a right R-module if we define $r' \cdot r = r'h(r)$ for all $r \in R$, $r' \in R'$.

2. $R' \otimes_R C$ then becomes a well-defined abelian group such that $r' \otimes rx = r'h(r) \otimes x$ for all $(r', r, x) \in R' \times R \times C$. We denote $R' \otimes_R C = R' \otimes_h C$.

3. $R' \otimes_h C$ becomes a left R'-module if we define $\rho(r' \otimes x) = \rho r' \otimes x$ for all $(\rho, r', x) \in R' \times R' \times C$).

4. If $f: C \to D$ is a homomorphism in the category of R-modules then $1 \otimes f: R' \otimes_h C \to R' \otimes_h D$ is a homomorphism in the category of R'-modules.

5. $(R' \otimes_h C, 1 \otimes d)$ is a chain complex over R'.

6. If $c = \{c^i_k\}$ is a basis of C then $\bar{c} = \{1 \otimes c^i_k\}$ is a basis of $R' \otimes_h C$. If b is another basis of C giving rise similarly to the basis $\bar{b}$ of $R' \otimes_h C$ then $\langle \bar{c}/\bar{b} \rangle = h_*(\langle c/b \rangle)$. If (a_{kj}) is the matrix of d_i with respect to c then $(h(a_{kj}))$ is the matrix of $1 \otimes d_i$ with respect to $\bar{c}$.

7. There is a simple isomorphism of (R', G') complexes, $R' \otimes_h C \cong C_h$ which takes the basis elements $1 \otimes c^i_k$ onto the basis element $\hat{c}^i_k$.

Thus we may and do identify: $R' \otimes_h C \equiv C_h$.

8. If $0 \to C' \xrightarrow{\alpha} C \xrightarrow{\beta} C'' \to 0$ is a split exact sequence of (R, G)-complexes with preferred bases c', c and c'' such that $c = \alpha(c') \cup b$ where $\beta(b) = c''$ then

$$0 \to C'_h = R' \otimes_h C' \xrightarrow{1 \otimes \alpha} C_h = R' \otimes_h C \xrightarrow{1 \otimes \beta} C''_h = R' \otimes_h C'' \to 0$$

is an exact sequence of (R', G')-complexes whose preferred bases have the analogous property.

Chapter IV

Whitehead Torsion in the CW Category

§19. The torsion of a CW pair—definition

The geometry in Chapter II and the algebraic analysis of Chapter III are synthesized in the definition:

If (K, L) *is a pair of finite, connected CW complexes such that* $K \searrow L$ *then the torsion of* (K, L)*—written* $\tau(K, L)$*—is defined by*

$$\tau(K, L) = \tau(C(\tilde{K}, \tilde{L})) \in Wh(\pi_1 L)$$

where $(\tilde{K}, \tilde{L})$ *is the universal covering of* (K, L).

In this section we explain this definition, show that $\tau(K, L)$ is well-defined and extend the definition to non-connected complexes. In the rest of the chapter we develop the basic properties of torsion in the CW category. This development allows us on the one hand (§24) to answer the questions about the relationship between homotopy type and simple-homotopy type which initiated our discussion. On the other hand, (§25), it yields the results on which the modern applications of simple-homotopy type are based.

In this chapter, as usual, all CW complexes mentioned, except those which arise as covering spaces, will be assumed to be finite.

Starting from the beginning, suppose that (K, L) is a connected CW pair and that $K \searrow L$. Let $p: \tilde{K} \to K$ be a universal covering, and let $G = \mathrm{Cov}(\tilde{K})$, the group of covering homeomorphisms. Then $p^{-1}L = \tilde{L}$ is a universal covering space of L and $\tilde{K} \searrow \tilde{L}$, by (3.13). The cellular chain complex $C(\tilde{K}, \tilde{L})$ is a $\mathbb{Z}(G)$-complex (see page 11). If we choose, for each cell $e_\alpha \in K - L$, a characteristic map φ_α and a specific lift $\tilde{\varphi}_\alpha$ of φ_α, then by (3.15) $B = \{\langle\tilde{\varphi}_\alpha\rangle \mid e_\alpha \in K - L\}$ is a basis for $C(\tilde{K}, \tilde{L})$ as a $\mathbb{Z}(G)$-complex. Let $\mathscr{B}$ be the set of all bases constructed in this fashion.

(19.1) *The complex* $C(\tilde{K}, \tilde{L})$, *along with the family of bases* $\mathscr{B}$, *determines an acyclic* $Wh(G)$*-complex.*

PROOF: $C(\tilde{K}, \tilde{L})$ is acyclic because $\tilde{K} \searrow \tilde{L}$ so, by (3.8), $H(C(\tilde{K}, \tilde{L})) \cong H(|\tilde{K}|, |\tilde{L}|) = 0$.

Suppose that $c, c' \in \mathscr{B}$ restrict to bases $c_n = \{\langle\tilde{\varphi}_1\rangle, \ldots, \langle\tilde{\varphi}_q\rangle\}$ and $c_n' = \{\langle\tilde{\psi}_1\rangle, \ldots, \langle\tilde{\psi}_q\rangle\}$ of $C_n(\tilde{K}, \tilde{L})$. Then

$$\langle\tilde{\psi}_j\rangle = \sum_k a_{jk}\langle\tilde{\varphi}_k\rangle \qquad \text{for some} \quad a_{jk} = \sum_i n_i^{jk} g_i \in \mathbb{Z}(G)$$

$$= \sum_{i,k} n_i^{j,k}\langle g_i\tilde{\varphi}_k\rangle.$$

But the cell $\tilde{\psi}_j(\mathring{I}^n)$, as a lift of e_j, is equal to one of the cells $g_{i_j}\tilde{\varphi}_j(\mathring{I}^n)$ and is disjoint from all of the others. Thus, by (3.7C), the coefficients in the last sum are all 0 except for $N = n_{i_j}^{j,j}$. But then $\tilde{\varphi}_j(\mathring{I}^n) = g_{i_j}^{-1}\tilde{\psi}_j(\mathring{I}^n)$ so, by the same argument, $\langle\tilde{\varphi}_j\rangle = N'\langle g_{i_j}^{-1}\tilde{\psi}_j\rangle$. Hence $\langle\tilde{\psi}_j\rangle = NN'\langle\tilde{\psi}_j\rangle$. So $N = \pm 1$ and $\langle\tilde{\psi}_j\rangle = \pm g_{i_j}\langle\tilde{\varphi}_j\rangle$. Therefore

$$\langle c_n/c_n'\rangle = \begin{pmatrix} \pm g_{i_1} & & 0 \\ & \ddots & \\ 0 & & \pm g_{i_q} \end{pmatrix}$$

and $\tau(\langle c_n/c_n'\rangle) = 0 \in Wh(G)$.

Thus $C(\tilde{K},\tilde{L})$ becomes a $Wh(G)$ complex if we stipulate that b is a preferred basis iff $\tau\langle c/b\rangle = 0$ for all $c \in \mathscr{B}$. □

Recall from §10, that there is a covariant functor which takes every group to its Whitehead group and every group homomorphism $G_1 \to G_2$ to a naturally induced homomorphism $Wh(G_1) \to Wh(G_2)$. In particular we now consider the induced isomorphisms $Wh(\pi_1(X, x)) \to Wh(\pi_1(X, y))$ corresponding to the change of base-point isomorphisms $\pi_1(X, x) \to \pi_1(X, y)$.

(19.2) *If X is an arcwise connected space containing the points x and y then all of the paths from x to y induce the same isomorphism $f_{x,y}$ of $Wh(\pi_1(X, x))$ onto $Wh(\pi_1(X, y))$. Moreover $f_{y,z} \circ f_{x,y} = f_{x,z}$.*

PROOF: If $\alpha:(I, 0, 1) \to (X, x, y)$, let $f_\alpha:\pi_1(X, x) \to \pi_1(X, y)$ denote the usual isomorphism given by $f_\alpha[\omega] = [\bar{\alpha} * \omega * \alpha]$. Then, if α, β are two such paths, $f_\beta^{-1}f_\alpha([\omega]) = [\beta*\bar{\alpha}]\cdot[\omega]\cdot[\beta*\bar{\alpha}]^{-1}$ for all $[\omega] \in \pi_1(X, x)$. Hence $f_\beta^{-1}f_\alpha$ is an inner automorphism and, by (10.4), $(f_\beta)_*^{-1}(f_\alpha)_* = (f_\beta^{-1}f_\alpha)_* = 1$. Thus $(f_\alpha)_* = (f_\beta)_*$ for all such α, β and we may set $f_{x,y} = (f_\alpha)_*$. It is obvious that $f_{y,z} \circ f_{x,y} = f_{x,z}$. □

Suppose as before that $p:\tilde{K} \to K$ is a universal covering, with $G = \mathrm{Cov}(\tilde{K})$ and K connected. Choosing base points $x \in K$ and $\tilde{x} \in p^{-1}(x)$ there is (page 12) an isomorphism $\theta = \theta(x, \tilde{x}):\pi_1(K, x) \to G$, given—if† we denote $\theta([\alpha]) = \theta_{[\alpha]}$, for all $[\alpha] \in \pi_1(K, x)$—by

$$\theta_{[\alpha]}(y) = \widetilde{\alpha * p\omega}(1),$$

where $y \in \tilde{K}$, ω is a path from $\tilde{x}$ to y, and $\widetilde{\alpha * p\omega}(0) = \tilde{x}$. If we identify $\pi_1(K, x)$ with G via θ then, by (19.1), $C(\tilde{K},\tilde{L})$ is an acyclic $Wh(\pi_1(K, x))$ complex and we may define $\tau(K, L) \in Wh(\pi_1(K, x))$.

To make the last sentence more precise (something worth doing only at the outset when we are worried about foundational questions) the isomorphism $\psi = \theta^{-1}:G \to \pi_1(K, x)$ induces a ring isomorphism of the same name, $\psi:\mathbb{Z}(G) \to \mathbb{Z}(\pi_1(K, x))$, and we wish to change rings as in §18 to construct from $C(\tilde{K},\tilde{L})$ the $Wh(\pi_1(K, x)$-complex $C(\tilde{K},\tilde{L})_\psi$. That $\tau(K, L)$ is independent of all choices will follow from

(19.3) *Let $\tilde{p}:\tilde{K} \to K$ and $\hat{p}:\hat{K} \to K$ be universal coverings of the connected complex K, with $\tilde{G}$ and $\hat{G}$ as the groups of covering homeomorphisms. Let $x, y \in K$, $\tilde{x} \in \tilde{p}^{-1}(x)$ and $\hat{y} \in \hat{p}^{-1}(y)$. Let $\tilde{\psi}:\tilde{G} \to \pi_1(K, x)$ and $\hat{\psi}:\hat{G} \to \pi_1(K, y)$*

†Thus $\theta_{[\alpha]}$ is the same as $g_{[\alpha]}$ of page 12.

be the group isomorphisms determined by $(x, \tilde{x})$ and $(y, \hat{y})$. Then $\tau(C(\hat{K}, \hat{L})_{\hat{\psi}}) = f_{x,y}\tau(C(\tilde{K}, \tilde{L})_{\tilde{\psi}})$

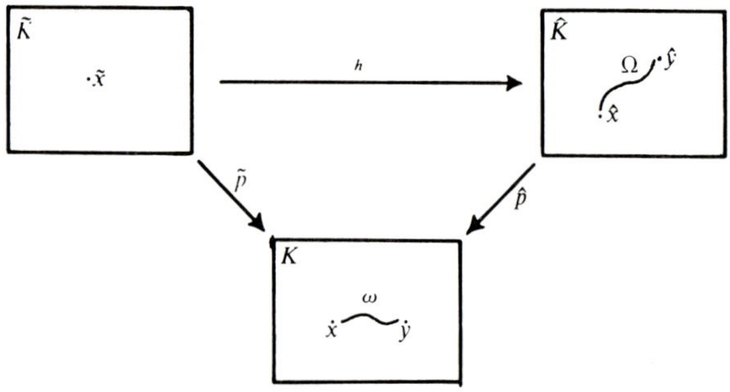

PROOF: Let $h:\tilde{K} \to \hat{K}$ be a homeomorphism covering the identity (hence a cellular isomorphism) and let $H:\tilde{G} \to \hat{G}$ by $H(g) = hgh^{-1}$. We claim first that $\tau C(\hat{K}, \hat{L}) = H_*(\tau C(\tilde{K}, \tilde{L}))$, where $H_*: Wh(\tilde{G}) \to Wh(\hat{G})$ is the induced map. To see this let $\{\langle\tilde{\varphi}_k^i\rangle\}$ be a basis for $C_i(\tilde{K}, \tilde{L})$ as in (19.1) and let $\hat{\varphi}_k^i = h \circ \tilde{\varphi}_k^i$ for all i, k. Then, by (19.1), the torsions of $C(\tilde{K}, \tilde{L})$ and $C(\hat{K}, \hat{L})$ can be computed using these bases. But, since h induces a chain isomorphism $C(\tilde{K}, \tilde{L}) \to C(\hat{K}, \hat{L})$ when these are thought of as complexes over $\mathbb{Z}$, one can check immediately that if the matrix of the boundary operator $\tilde{d}_i$ is $(a_{kj}) \in GL(\mathbb{Z}\tilde{G})$ then the matrix of the corresponding boundary operator $\hat{d}_i$ is $(H(a_{kj}))$, where $H:\mathbb{Z}(\tilde{G}) \to \mathbb{Z}(\hat{G})$ is induced from $H:\tilde{G} \to \hat{G}$. Thus, by the proof of (18.2), $\tau C(\hat{K}, \hat{L}) = H_*(\tau C(\tilde{K}, \tilde{L}))$.

Now let $\hat{x} = h(\tilde{x})$ and choose a path $\Omega:(I, 0, 1) \to (\hat{K}, \hat{x}, \hat{y})$. Let $\omega = \hat{p}\Omega$ and let $f_\omega: \pi_1(K, x) \to \pi_1(K, y)$ as usual. Then denoting $\tilde{\theta} = \theta(x, \tilde{x}) = \tilde{\psi}^{-1}$ and $\hat{\theta} = \theta(y, \hat{y}) = \hat{\psi}^{-1}$, the following diagram commutes

$$\begin{array}{ccc} \tilde{G} & \xrightarrow{\quad H \quad} & \hat{G} \\ \Big\uparrow{\scriptstyle\tilde{\theta}} & & \Big\uparrow{\scriptstyle\hat{\theta}} \\ \pi_1(K, x) & \xrightarrow{\quad f_\omega \quad} & \pi_1(K, y) \end{array}$$

For if $[\alpha] \in \pi_1(K, x)$ we have

$$\begin{aligned} (\hat{\theta} f_\omega[\alpha])(\hat{y}) &= \hat{\theta}_{[\bar{\omega}*\alpha*\omega]}(\hat{y}) \\ &= \widehat{\alpha*\omega}(1) \quad \text{where} \quad \widehat{\alpha*\omega}(0) = \bar{\Omega}(1) = \hat{x} \\ &= h(\widetilde{\alpha*\omega}(1)) \quad \text{because} \quad h(\widetilde{\alpha*\omega})(0) = \hat{x} \\ &= h(\widetilde{\alpha*\tilde{p}h^{-1}\Omega}(1)) \quad \text{because} \quad \omega = \hat{p}\Omega = \tilde{p}h^{-1}\Omega \\ &= h(\tilde{\theta}_{[\alpha]}h^{-1}\Omega(1)) \\ &= (h\tilde{\theta}_{[\alpha]}h^{-1})(\hat{y}) \\ &= (H\tilde{\theta}[\alpha])(\hat{y}). \end{aligned}$$

Since $\hat{\theta} f_\omega[\alpha]$ and $H\tilde{\theta}[\alpha]$ agree at a point they agree everywhere. Since $[\alpha]$ was arbitrary, $\hat{\theta} f_\omega = H\tilde{\theta}$.

From the preceding paragraphs and (18.2) we have,

$$\begin{aligned} f_{x,y}\tau(C(\tilde{K},\tilde{L})_{\tilde{\psi}} &= f_{\omega *}\tilde{\psi}_*(\tau C(\tilde{K},\tilde{L})) \\ &= \hat{\psi}_* H_*(\tau C(\tilde{K},\tilde{L}) \\ &= \hat{\psi}_*(\tau C(\hat{K},\hat{L})) \\ &= \tau(C(\hat{K},\hat{L})_{\hat{\psi}}). \quad \square \end{aligned}$$

Since it would sometimes be nice to have $\tau(K, L)$ defined as a single element of a single group we introduce some formalism. Let us (using (19.2)) define

$$Wh(\pi_1 K) = [\bigcup_{x\in K} Wh(\pi_1(K, x))]/''\sim''$$

where $a \sim b$ if $a \in Wh(\pi_1(K, x))$, $b \in Wh(\pi_1(K,y))$ and $f_{x,y}(a) = b$. Let $j_x : Wh(\pi_1(K, x)) \to Wh(\pi_1 K)$ be the natural bijection. The stipulation that j_x be a group isomorphism gives $Wh(\pi_1 K)$ a group structure which is independent of x. Note that $j_y^{-1} j_x = f_{x,y}$.

If $f:(K, x) \to (K', x')$, then the induced homomorphism on fundamental groups gives rise to a composite homomorphism

$$f_* : Wh(\pi_1 K) \xrightarrow{j_x^{-1}} Wh(\pi_1(K, x)) \xrightarrow{f_\#} Wh(\pi_1(K', x')) \xrightarrow{j_{x'}} Wh(\pi_1 K')$$

The proof of the following is left to the reader.

(19.4) *The homomorphism f_* is independent of which pair (x, x') with $f(x) = x'$ is chosen. Thus there is a covariant functor from the category of finite connected CW complexes and maps to the category of abelian groups and homomorphisms defined by*

$$K \mapsto Wh(\pi_1 K)$$

$$\{f: K \to K'\} \mapsto \{f_* : Wh(\pi_1 K) \to Wh(\pi_1 K')\}.$$

Moreover, if $f \simeq g$ then $f_* = g_*$. $\square$

Putting all this together, $\tau(K, L) \in Wh(\pi_1 L)$ is defined (in the connected case) as follows: Choose a point $x \in K$, a universal covering $p:(\tilde{K},\tilde{L}) \to (K, L)$ and a point $\tilde{x} \in p^{-1}(x)$. Let $i: L \to K$ be the inclusion. Then $\tau(K, L)$ is the end of the sequence

$$\begin{array}{c} \tau C(\tilde{K},\tilde{L}) \in Wh(G) \\ \downarrow \psi(x,\tilde{x})_* \\ \tau(C(\tilde{K},\tilde{L}) \in Wh(\pi_1(K, x)) \\ \downarrow j_x \\ \tau' \in Wh(\pi_1 K) \\ \downarrow i_*^{-1} \\ \tau(K, L) \in Wh(\pi_1 L) \end{array}$$

$\tau(K, L)$ is well-defined by (19.1)–(19.4).

The reader may wonder why we don't omit i_*^{-1} and put $\tau(K, L)$ into $Wh(\pi_1 K)$ instead of $Wh(\pi_1 L)$. It's a matter of taste. The discussion in §6 seems to lend weight to the view that L is the central object in our discussion.

Non-Connected Case

Finally, we generalize to the non-connected case. Assume that K and L are finite CW complexes and that $K \searrow\!\!\!\nearrow L$. Let $K_1, \ldots K_q$, and $L_1, \ldots, L_q$ be the components of K and L respectively, ordered so that $K_j \searrow\!\!\!\nearrow L_j$ for all j. We define

$$\tau(K, L) = \sum_j \tau(K_j, L_j) \in \oplus\, Wh(\pi_1 L_j)$$

In §21 we shall justify, and thereafter we shall use, the notational convention

$$Wh(L) \equiv \oplus\, Wh(\pi_1 L_j).$$

Note that (19.4) generalizes to

(19.5) *There is a covariant functor from the category of finite CW complexes and maps to the category of abelian groups and homomorphisms defined by*

$$K \mapsto \bigoplus_{j=1}^{q} Wh(\pi_1 K_j)$$

$$\{f: K \to K'\} \to \{f_* = \sum_{j=1}^{q} f_{j*}: \bigoplus_{j=1}^{q} Wh(\pi_1 K_j) \to \bigoplus_{i=1}^{r} Wh(\pi_1 K_i')\}$$

where $K_1, \ldots, K_q$ and $K_1', \ldots, K_r'$ are the components of K and K' respectively, and $f_{j*}: Wh(\pi_1 K_j) \to Wh(\pi_1 K_{i_j}')$ is induced from f with $f(K_j) \subset K_{i_j}'$. Moreover, if $f \simeq g$ then $f_* = g_*$. □

Two comments are in order:

First, the reader must NOT confuse $\oplus Wh(\pi_1 K_j)$ with $Wh(\oplus \pi_1 K_j)$. For example, (11.5) implies that $Wh(\mathbb{Z}_3) \oplus Wh(\mathbb{Z}_4) \neq Wh(\mathbb{Z}_{12})$. Torsion considerations are first done for each component K_j, and then formally added.

Second, despite the first comment, it is not a sterile generalization to consider the non-connected case. Sometimes connected spaces are expressed as the union of non-connected spaces, or as the union of connected spaces along a non-connected intersection. The Excision Lemma (20.3) and the Sum Theorem (23.1) would be much less useful if the theory were developed with the connectivity restrictions. The point is that formal addition becomes real addition under f_* if f carries different components into the same component.

Having rigorously defined $\tau(K, L)$ we can allow ourselves some laxity in the ensuing discussion. Thus, for sake of clarity, we shall (when K is connected) sometimes speak of $\tau(K, L)$ as an element of $Wh(G)$, or as an element of $Wh(\pi_1(L, x))$, for some $x \in L$. At other times (also in the name of clarity) we shall be completely rigorous and consider $\tau(K, L)$ as an element of $Wh(\pi_1 L)$.

§20. Fundamental properties of the torsion of a pair

(20.1) *If (K, L) is a CW pair such that $K \searrow L$ and if each component of $K-L$ is simply connected then $\tau(K, L) = 0$.*

PROOF: Clearly it suffices to prove this when K is connected. Let c be a component of $K-L$. Then c is closed in $K-L$, so $\bar{c} \subset L \cup c$ and $L \cup c$ is a closed set. If e is a cell of K which meets c then e cannot lie totally in L, so, L being a subcomplex, $e \cap L = \emptyset$. Hence $e \subset K-L$ and consequently $e \subset c$. Combining these facts we see that $L \cup c$ is a subcomplex of K and $c = (L \cup c)-L$ is a union of cells.

As usual let $p:\tilde{K} \to K$ be a universal covering with G the group of covering homeomorphisms. Since c is simply connected it lifts homeomorphically to $\tilde{K}$. Let C be one lift of c, so $p|C:C \to c$ is a homeomorphism. Let $\{gC \mid 1 \neq g \in G\}$ be the other lifts. These lifts are pairwise disjoint since c is connected. For each p-cell e_α of c let $\tilde{\varphi}_\alpha$ be a lift of a characteristic map such that $\tilde{\varphi}_\alpha(\mathring{I}^p) \subset C$. Doing this for all components c of $K-L$, and all such cells e_α, we get a preferred basis $\{\langle\tilde{\varphi}_\alpha\rangle\}$ for $C(\tilde{K}, \tilde{L})$ (which we are thinking of as a $Wh(G)$-complex).

For a fixed n-cell e_α of the component c of $K-L$,

$$\partial\langle\tilde{\varphi}_\alpha\rangle \in H_{n-1}(\tilde{K}^{n-1} \cup \tilde{L}, \tilde{K}^{n-2} \cup \tilde{L})$$

is represented by a singular cycle carried by $\tilde{\varphi}_\alpha(\partial I^n)$. However $\tilde{\varphi}_\alpha(\mathring{I}^n) \subset C$ and $\varphi_\alpha(I^n) \subset L \cup c$, so $\tilde{\varphi}_\alpha(\partial I^n) \subset \tilde{L} \cup C$. Thus any $(n-1)$-cell of $\tilde{K}-\tilde{L}$ which meets $\tilde{\varphi}_\alpha(\partial I^n)$ must lie in C. It follows from (3.7c) that in the expression

$$\partial\langle\tilde{\varphi}_\alpha\rangle = \sum_{\beta, j} n_{\alpha\beta j} g_j \langle\tilde{\varphi}_\beta\rangle = \sum n_{\alpha\beta j}\langle g_j\tilde{\varphi}_\beta\rangle, \quad (n_{\alpha\beta j} \in \mathbb{Z}, g_j \in G)$$

we must have $n_{\alpha\beta j} = 0$ unless $g_j\tilde{\varphi}_\beta(\mathring{I}^{n-1}) \subset C$. But, by choice of our preferred basis $g_j\tilde{\varphi}_\beta(\mathring{I}^{n-1}) \subset C$ only if $g_j = 1$. Thus

$$\partial\langle\varphi_\alpha\rangle = \sum_\beta n_{\alpha\beta}\langle\tilde{\varphi}_\beta\rangle$$

and we see that the matrix of ∂ has only integer entries. Thus, by (18.3), $\tau(C(\tilde{K}, \tilde{L})) = 0 \in Wh(G)$. □

(20.2) *If $K > L > M$ where $K \searrow L$ and $L \searrow M$ then*

$$\tau(K, M) = \tau(L, M) + i_*^{-1}\tau(K, L)$$

where $i: M \subsetneq L$.

PROOF: We may assume K is connected. Let $p:\tilde{K} \to K$ be the universal covering. Set $\tilde{L} = p^{-1}L$, $G' = \text{Cov}(\tilde{L})$ and $G = \text{Cov}(\tilde{K})$. If $j:L \subsetneq K$ then (page 12) $j_\#:G' \to G$ is an isomorphism. Note [using (3.16) with $\tilde{j}:\tilde{L} \subsetneq \tilde{K}$] that $j_\#(g') \in G$ is the unique extension of g'. Set $J = j_\#$ and also let J denote the induced map $\mathbb{Z}(G') \xrightarrow{\cong} \mathbb{Z}(G)$.

By (19.1), $C(\tilde{K}, \tilde{M})$ and $C(\tilde{K}, \tilde{L})$ are $Wh(G)$-complexes and $C(\tilde{L}, \tilde{M})$ is a

$Wh(G')$-complex. But $C(\tilde{L}, \tilde{M})$ may also be viewed as a $Wh(G)$-complex if, given $g \in G$ and $x \in C(\tilde{L}, \tilde{M})$, we define $g \cdot x = g' \cdot (x)$ where $J(g') = g$. Choose a preferred basis $\{\langle \tilde{\varphi}_\alpha \rangle | e_\alpha \in K - M\}$ for $C(\tilde{K}, \tilde{M})$ as a $\mathbb{Z}(G)$-module and use the same lifts $\tilde{\varphi}_\alpha$ to give preferred bases $\{\langle \tilde{\varphi}_\alpha \rangle | e_\alpha \in K - L\}$ and $\{\langle \tilde{\varphi}_\alpha \rangle | e_\alpha \in L - M\}$ to the $\mathbb{Z}$(G)-modules $C(\tilde{K}, \tilde{L})$ and $C(\tilde{L}, \tilde{M})$. Then the inclusion maps induce a short exact sequence

$$0 \to C(\tilde{L}, \tilde{M}) \to C(\tilde{K}, \tilde{M}) \to C(\tilde{K}, \tilde{L}) \to 0$$

of acyclic $Wh(G)$-complexes in which preferred bases correspond. Hence, by (17.2), $\tau C(\tilde{K}, \tilde{M}) = \tau C(\tilde{L}, \tilde{M}) + \tau C(\tilde{K}, \tilde{L})$.

Now think of $C(\tilde{L}, \tilde{M})$ as a $Wh(G')$-complex and notice that, by definition of $C(\tilde{L}, \tilde{M})_J$ there is a trivial basis preserving isomorphism of $C(\tilde{L}, \tilde{M})_J$ with the complex "$C(\tilde{L}, \tilde{M})$ viewed as a $Wh(G)$-complex" discussed above. Thus the torsion of the latter complex is equal to $\tau(C(\tilde{L}, \tilde{M})_J) = J_* \tau C(\tilde{L}, \tilde{M})$. Hence $\tau C(\tilde{K}, \tilde{M}) = [\tau C(\tilde{K}, \tilde{L}) + J_* \tau C(\tilde{L}, \tilde{M})] \in Wh(G)$. The theorem now follows immediately if one traces each term in this equation to its image in $Wh(\pi_1 M)$ via the following commutative diagram

$$\begin{array}{ccc} Wh(G') & \xrightarrow{J_*} & Wh(G) \\ \downarrow & & \downarrow \\ Wh(\pi_1 L) & \xrightarrow{j_*} & Wh(\pi_1 K) \\ & {\scriptstyle i_*}\nwarrow \quad \nearrow{\scriptstyle k_*} & \\ & Wh(\pi_1 M) & \end{array}$$

Here i, j, k are inclusions and the vertical arrows are the result of the discussion in §19. The commutativity of the diagram is left as an exercise for the reader. □

(20.3) (The Excision Lemma) *If K, L, and M are subcomplexes of the complex $K \cup L$, with $M = K \cap L$, and if $K \nearrow\!\!\searrow M$ then $\tau(K \cup L, L) = j_* \tau(K, M)$ where $j: M \to L$ is the inclusion map.*

PROOF: First, we claim, it suffices to prove this when L is connected. (This does not say that K and M are connected.) For suppose that L and $K \cup L$ have components $L_1, \ldots, L_q$ and $P_1, \ldots, P_q$ where $P_i \nearrow\!\!\searrow L_i$ for all i. Let $M_i = M \cap L_i$ and $K_i = K \cap P_i$ have components $M_{i1}, M_{i2}, \ldots$ and $K_{i1}, K_{i2}, \ldots$ respectively, where $K_{ik} \nearrow\!\!\searrow M_{ik}$. Then, assuming the Excision Lemma for each L_i, we have

$$\begin{aligned} \tau(K \cup L, L) &\equiv \sum_i \tau(P_i, L_i) \\ &= \sum_i j_{i*} \tau(K_i, M_i) \quad \text{where} \quad j_i : M_i \subsetneq L_i \\ &= \sum_i j_{i*} \Big(\sum_k \tau(K_{ik}, M_{ik}) \Big) \\ &= \sum_{i,k} (j_i | M_{ik})_* \tau(K_{ik}, M_{ik}) \\ &\equiv j_* \tau(K, M). \end{aligned}$$

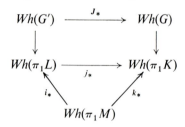

So assume that L (hence also $K \cup L$) is connected and that M and K have components $M_1, \ldots, M_s$, and $K_1, \ldots, K_s$ with $K_i \searrow M_i$. The proof of the theorem consists of a technical rendering of the fact that $(K \cup L) - L = \bigcup_i (K_i - M_i)$ where the $K_i - M_i$ are disjoint.

Let $p: \widetilde{K \cup L} \to K \cup L$ be a universal covering of $K \cup L$ and denote $\tilde{X} = p^{-1}X$ if $X \subset K \cup L$. [Beware: In general $\tilde{K}_i = p^{-1}K_i$ is not a universal covering of K_i.] Let $G = \operatorname{Cov}(\widetilde{K \cup L})$. Note that $C(\widetilde{K \cup L}, \tilde{L}) = \bigoplus_i C(\widetilde{K_i \cup L}, \tilde{L})$, where all the chain complexes in this equation can be viewed as acyclic $Wh(G)$-complexes. Hence, by (17.1),

$$\tau(K \cup L, M) = \sum_i \tau C(\widetilde{K_i \cup L}, \tilde{L}) \in Wh(G).$$

To compute $\tau C(\widetilde{K_i \cup L}, \tilde{L})$ we consider a universal covering $\hat{p}: (\hat{K}_i, \hat{M}_i) \to (K_i, M_i)$ with $\hat{G}_i = \operatorname{Cov}(\hat{K}_i)$. Fixing base points $x \in M_i$, $\hat{x} \in \hat{p}^{-1}(x)$, and $\tilde{x} \in p^{-1}(x)$ and letting $J_i: (K_i, x) \to (K \cup L, x)$ be the inclusion, the following commutative diagrams are determined:

$$\begin{array}{ccc} (\hat{K}_i, \hat{x}) & \xrightarrow{\tilde{J}_i} & (\widetilde{K \cup L}, \tilde{x}) \\ \downarrow{\scriptstyle \hat{p}} & & \downarrow{\scriptstyle p} \\ (K_i, x) & \xrightarrow[J_i]{} & (K \cup L, x) \end{array} \qquad \begin{array}{ccc} \hat{G}_i & \xrightarrow{\lambda} & G \\ \downarrow{\scriptstyle \cong} & & \downarrow{\scriptstyle \cong} \\ \pi_1(K_i, x) & \xrightarrow{J_{i\#}} & \pi_1(K \cup L, x) \\ \uparrow{\scriptstyle \cong\ \text{incl.}} & & \uparrow{\scriptstyle \cong\ \text{incl.}} \\ \pi_1(M_i, x) & \xrightarrow{(j|M_i)_\#} & \pi_1(L, x) \end{array}$$

We shall use the fact (3.16) that $\tilde{J}_i \circ g = \lambda(g) \circ \tilde{J}_i$, if $g \in \hat{G}$. (In the notation of (3.16) the map λ should also be denoted by $J_{i\#}$.)

Now for each cell $e_\alpha \in K_i - M_i$ with characteristic map φ_α choose a fixed lift $\hat{\varphi}_\alpha$ to $\hat{K}_i$ and define $\tilde{\varphi}_\alpha = \tilde{J}_i \circ \hat{\varphi}_\alpha$. Then $\{\langle\tilde{\varphi}_\alpha\rangle\}$ is a basis for the $Wh(G)$-complex $C(\widetilde{K_i \cup L}, \tilde{L})$. Also $\tilde{J}_i$ induces a chain map (over $\mathbb{Z}$), since it is cellular. Hence if $\partial\langle\hat{\varphi}_\alpha\rangle = \sum n_{\alpha\beta\gamma} g_\gamma \langle\hat{\varphi}_\beta\rangle$, where $g_\gamma \in \hat{G}_i$, we have,

$$\begin{aligned} \partial\langle\tilde{\varphi}_\alpha\rangle &= \tilde{J}_{i*}\partial\langle\hat{\varphi}_\alpha\rangle \\ &= \sum n_{\alpha\beta\gamma}(\tilde{J}_i \circ g_\gamma)_*\langle\hat{\varphi}_\beta\rangle \\ &= \sum n_{\alpha\beta\gamma}(\lambda(g_\gamma) \circ \tilde{J}_i)_*\langle\hat{\varphi}_\beta\rangle \\ &= \sum n_{\alpha\beta\gamma}\lambda(g_\gamma)\langle\tilde{\varphi}_\beta\rangle. \end{aligned}$$

Hence $C(\widetilde{K_i \cup L}, \tilde{L})$ is simply isomorphic to $C(\hat{K}_i, \hat{M}_i)_\lambda$. So $\tau C(\widetilde{K_i \cup L}, \tilde{L}) = \lambda_* \tau C(\hat{K}_i, \hat{M}_i) \in Wh(G)$. This corresponds, by the right-hand diagram above, to $(j|M_i)_*\tau(K_i, M_i) \in Wh(\pi_1 L)$. Thus

$$\tau(K \cup L, L) = \sum_i (j|M_i)_*\tau(K_i, M_i) \equiv j_*\tau(K, M). \quad \square$$

As an immediate consequence of (20.2) and (20.3) we get

(20.4) *If K, L and M are subcomplexes of the complex $K \cup L$, with $M = K \cap L$ and if $K \searrow M$ and $L \searrow M$ then $\tau(K \cup L, M) = \tau(K, M) + \tau(L, M)$.* □

(20.5) *Suppose that (K, L) is a connected CW pair in simplified form* (see page 26), $K = L \cup \bigcup_j e_j^n \cup \bigcup_i e_i^{n+1}$ $(n \geq 2)$, *and that $\{\psi_j\}$ and $\{\varphi_i\}$ are characteristic maps for the e_j^n and e_i^{n+1}. Set $K_n = L \cup \bigcup_j e_j^n$. Let $\langle\partial\rangle$ be the matrix—with entries in $\mathbb{Z}(\pi_1(L, e^0))$—of the boundary operator $\partial: \pi_{n+1}(K, K_n; e^0) \to \pi_n(K_n, L; e^0)$ with respect to the bases $\{[\psi_j]\}$ and $\{[\varphi_i]\}$ given by* (8.1). *Then $\tau(K, L) = (-1)^n \tau\langle\partial\rangle$.*

PROOF: It follows from the proof of (8.1) and the fact that the Hurewicz map commutes with boundary operators that there is a commutative diagram

$$
\begin{array}{ccc}
0 & & \\
\downarrow & & \\
H_{n+1}(\tilde{K}, \tilde{K}_n) & \xrightarrow{\cong} & \pi_{n+1}(K, K_n; e^0) \\
\downarrow \partial & & \downarrow \partial \\
H_n(\tilde{K}_n, \tilde{L}) & \xrightarrow{\cong} & \pi_n(K_n, L; e^0) \\
\downarrow & & \\
0 & &
\end{array}
$$

in which the preferred bases $\{\langle\tilde{\varphi}_i\rangle\}$ and $\{\langle\tilde{\psi}_j\rangle\}$ go to the bases $\{[\varphi_i]\}$ and $\{[\psi_j]\}$. Since the left-hand column is just $C(\tilde{K}, \tilde{L})$, the result follows from P3 of (17.1). □

§21. The natural equivalence of *Wh(L)* and $\oplus$ *Wh($\pi_1 L_j$)*

We have considered two functors from the category of finite CW complexes and maps to the category of abelian groups and group homomorphisms. In §6 we defined the functor

$$L \mapsto Wh(L)$$
$$\{f: L \to L'\} \mapsto \{f_*: Wh(L) \to Wh(L')\}$$

and in §19 we defined (19.5) the functor

$$L \to \bigoplus_j Wh(\pi_1 L_j), \quad (L_1, \ldots, L_q \text{ the components of } L)$$
$$\{f: L \to L'\} \mapsto \{f_* = \sum_j f_{j*}: \bigoplus_j Wh(\pi_1 L_j) \to \bigoplus_i Wh(\pi_1 L_i')\}.$$

(It will be up to the reader to keep the two meanings of f_* straight.) The purpose of this section is to prove

(21.1) *For every finite CW complex L define $T_L : Wh(L) \to \bigoplus_j Wh(\pi_1 L_j)$, where the L_j are the components of L, by $T_L([K, L]) = \tau(K, L)$. Then $T = \{T_L\}$ is a natural equivalence of functors.*

REMARK: After having proved this we shall adopt the habit of writing $Wh(L)$ for $\oplus\, Wh(\pi_1 L_j)$.

PROOF: For each L, T_L is well-defined: For if $L < K < K'$ where $K' \searrow^{e} K$, (recall: "$\searrow^{e}$" denotes an elementary collapse) then $K' \searrow K$ and $K' - K$ is simply connected. Thus, by (20.1) and (20.2)

$$\tau(K', L) = \tau(K, L) + i_*^{-1}\tau(K', K) = \tau(K, L).$$

By induction on the number of elementary collapses and expansions we see that $\tau(K, L) = \tau(K', L)$ if $K \nearrow\!\!\searrow K'$ rel L. Hence $T_L([K, L]) = T_L([K', L])$ if $[K, L] = [K', L]$.

For each L, T_L is a homomorphism: This is exactly the content of (20.4).

T_L is one-one: For suppose that $T_L([K, L]) = \tau(K, L) = 0$. We may assume, by (7.4) and the fact that T_L is well-defined, that each component of (K, L) is in simplified form. Hence, by (20.5), $\tau\langle \partial^j \rangle = 0$, where ∂^j is the usual boundary operator in homotopy for the jth component. But by (8.4) this implies that $K \nearrow\!\!\searrow L$ rel L. Thus $[K, L] = 0$ and, T_L is injective.

T_L is onto by (20.5) and (8.7).

Thus for each L, T_L is an isomorphism. To prove that T is a natural equivalence it remains to show that, if $f: L \to L'$, the following diagram commutes.

$$\begin{array}{ccc}
Wh(L) & \xrightarrow{\;f_*\;} & Wh(L') \\
\Big\downarrow{\scriptstyle T_L} & & \Big\downarrow{\scriptstyle T_{L'}} \\
\bigoplus_j Wh(\pi_1 L_j) & \xrightarrow{\;f_*\;} & \bigoplus_i Wh(\pi_1 L'_i)
\end{array}$$

We may assume that f is cellular. So, given $[K, L] \in Wh(L)$,

$$\begin{aligned}
T_{L'} f_*[K, L] &= \tau(K \cup_L M_f, L'), \text{ by def. of } f_* \text{ (page 22)} \\
\overset{(20.2)}{=}\;& \tau(M_f, L') + i_*^{-1}\tau(K \cup_L M_f, M_f), && \text{where } i: L' \subset M_f \\
&= i_*^{-1}\tau(K \cup_L M_f, M_f), && \text{since } M_f \searrow L' \\
&= p_*\tau(K \cup_L M_f, M_f), && \text{where } p: M_f \to L' \text{ is the natural projection} \\
\overset{\text{excision (20.3)}}{=}\;& p_* j_* \tau(K, L), && \text{where } j: L \overset{\subset}{\to} M_f \\
&= f_*\tau(K, L) && \text{since } pj = f. \quad \square
\end{aligned}$$

§22. The torsion of a homotopy equivalence

Suppose that $f: K \to L$ is a cellular homotopy equivalence between finite CW complexes. Then $M_f \searrow K$ and $f_*: Wh(K) \to Wh(L)$ is an isomorphism. We define

$$\tau(f) = f_*\tau(M_f, K) \in Wh(L).$$

In this section we give some formal properties of the torsion of a homotopy equivalence and show how it may sometimes be computed. We shall tacitly and frequently use the equivalence of §21.

(22.1) *If $f, g: K \to L$ are homotopic cellular homotopy equivalences then $\tau(f) = \tau(g)$.*

PROOF: $f_* = g_*$ by (19.5). Thus it suffices to show that $\tau(M_f, K) = \tau(M_g, K)$. This is true because $M_f \curlywedge M_g$ rel K, by (5.5). □

As a consequence of this lemma we may define $\tau(f)$ when $f: K \to L$ is an arbitrary homotopy equivalence by setting $\tau(f) = \tau(g)$, where g is any cellular approximation to f. Thus, while the propositions and proofs in this section are stated for cellular maps, one often thinks of them as propositions about arbitrary maps.

(22.2) *A cellular homotopy equivalence $f: K \to L$ is a simple-homotopy equivalence if and only if $\tau(f) = 0$.*

(Although this statement is a theorem for us—simple-homotopy equivalence having been defined geometrically in §4—it is often taken as the definition of simple-homotopy equivalence.)

PROOF: Since f_* is an isomorphism, $\tau(f) = 0$ iff $\tau(M_f, K) = 0$. But by (21.1), this is true iff $M_f \curlywedge K$ rel K. And that is true (5.8) iff f is a simple-homotopy equivalence. □

(22.3) *If $L < K$ and $K \searrow L$ then $\tau(i) = i_*\tau(K, L)$ where $i: L \to K$ is the inclusion map.*

PROOF: $M_i = (L \times I) \underset{L \times 1}{\cup} K \times 1$, where $L \equiv L \times 0 < M_i$. Then

$$M_i \nearrow (K \times I) \searrow K \times 0 \equiv K \text{ rel } L.$$

Hence $\tau(M_i, L) = \tau(K, L)$, so that $\tau(i) = i_*\tau(M_i, L) = i_*\tau(K, L)$. □

(22.4) *If $f: K \to L$ and $g: L \to M$ are cellular homotopy equivalences* then *$\tau(gf) = \tau(g) + g_*\tau(f)$.*

PROOF:
$$\begin{aligned}\tau(gf) &= g_* f_* \tau(M_{gf}, K)\\ &= g_* f_* \,[\tau(M_f \underset{L}{\cup} M_g, K)], \quad \text{by (5.6)}\\ &= g_* f_*[\tau(M_f, K) + i_*^{-1}\tau(M_f \underset{L}{\cup} M_g, M_f)],\end{aligned}$$

where $i: K \subsetneq M_f$, using (20.2)

$$= g_*\tau(f)+g_*f_*[i_*^{-1}j_*\tau(M_g, L)],$$

where $j:L \subsetneq M_f$, using "excision" (20.3),

$$= g_*\tau(f)+g_*\tau(M_g, L),$$

since $f = pi$ and $1_L = pj$ imply $f_*i_*^{-1}j_* = 1_{Wh(\pi_1 L)}$

$$= g_*\tau(f)+\tau(g). \quad \square$$

As a corollary we get

(22.5) *If $f:K \to L$ and $g:L \to K$ are cellular homotopy equivalences which are homotopy inverses of each other then $\tau(g) = -g_*\tau(f)$.*

PROOF: Since $gf \simeq 1_K$ we have $0 = \tau(gf) = \tau(g)+g_*\tau(f)$. $\square$

(22.6) *If $f:(K, K_0) \to (L, L_0)$ where $K \searrow K_0$, $L \searrow L_0$, and if $f:K \to L$ and $\bar{f} = f|K_0:K_0 \to L_0$ are cellular homotopy equivalences,*[13] *then*

(a) $\tau(f) = j_*\tau(\bar{f})+[\tau(j)-f_*\tau(i)]$

(b) $\tau(L, L_0) = \bar{f}_*\tau(K, K_0)+[D_*\tau(f)-\tau(\bar{f})]$

where $i:K_0 \subset K$, $j:L_0 \subset L$, and $D:L \to L_0$ is a deformation retraction.

PROOF: Clearly $fi = j\bar{f}$. Thus

$$\tau(f)+f_*\tau(i) = \tau(j)+j_*\tau(\bar{f}), \text{ proving (a).}$$

Further:

$$\tau(j) = f_*\tau(i)+\tau(f)-j_*\tau(\bar{f})$$

$$j_*\tau(L, L_0) = f_*i_*\tau(K, K_0)+\tau(f)-j_*\tau(\bar{f}), \text{ by (22.3)}$$

$$\tau(L, L_0) = (DFi)_*\tau(K, K_0)+D_*\tau(f)-\tau(\bar{f})$$

$$= \bar{f}_*\tau(K, K_0)+[D_*\tau(f)-\tau(\bar{f})],$$

proving (b). $\square$

As a corollary we get

(22.7) *If $f:(K, K_0) \to (L, L_0)$ as in* (22.6) *and if f and $\bar{f}$ are simple-homotopy equivalences, then* (a) $\tau(j) = f_*\tau(i)$ *and* (b) $\tau(L, L_0) = \bar{f}_*\tau(K, K_0)$. $\square$

The brute force calculation of torsion

To actually get down to the nuts and bolts of computing $\tau(f)$, one proceeds as follows.

Suppose that $f:K \to L$ is a cellular homotopy equivalence between connected spaces and that $\tilde{f}:\tilde{K} \to \tilde{L}$ is a lift of f to universal covering spaces inducing $\tilde{f}_*:C(\tilde{K}) \to C(\tilde{L})$. Let G_K and G_L be the groups of covering homeomorphisms of $\tilde{K}$ and $\tilde{L}$, and let $C(\tilde{K})$ and $C(\tilde{L})$ be viewed as $Wh(G_K)$-

---

[13] The hypothesis $L \searrow L_0$ is redundant.

and $Wh(G_L)$-complexes with boundary operators d and d' respectively. Choose base points, x, y, and points covering them $\tilde{x}$, $\tilde{y}$ such that $f(x) = y$ and $\tilde{f}(\tilde{x}) = \tilde{y}$. Let $f_\# : G_K \to G_L$ be induced from $f_\# : \pi_1(K, x) \to \pi_1(L, y)$ as in (3.16). [Also let $f_\#$ denote the corresponding maps $\mathbb{Z}G_K \to \mathbb{Z}G_L$ and $GL(\mathbb{Z}G_K) \to GL(\mathbb{Z}G_L)$]. In these circumstances we have

(22.8) $\tau(f) \in Wh(G_L)$ *is the torsion of the* $Wh(G_L)$-*complex* $\mathscr{C}$ *which is given by*

(1) $\mathscr{C}_n = [C(\tilde{K})_{f_\#}]_{n-1} \oplus C_n(\tilde{L})$

(2) $\partial_n : \mathscr{C}_n \to \mathscr{C}_{n-1}$ *has matrix*

$$
\begin{array}{cc}
 & [C(\tilde{K})_{f_\#}]_{n-2} \quad C_{n-1}(\tilde{L}) \\
\begin{array}{c} [C(\tilde{K})_{f_\#}]_{n-1} \\ \\ C_n(\tilde{L}) \end{array} &
\left(
\begin{array}{c|c}
-f_\#\langle d_{n-1}\rangle & \langle \tilde{f}_*\rangle \\
\hline
\bigcirc & \langle d'_n\rangle
\end{array}
\right)
\end{array}
$$

In particular if we let $\bar{C}(\tilde{K})$ *be the* $Wh(G_L)$-*complex with* $\bar{C}_n(\tilde{K}) = [C(\tilde{K})_{f_\#}]_{n-1}$ *and boundary operator* $\bar{d}$ *given by* $\langle \bar{d}_n \rangle = -f_\#\langle d_{n-1}\rangle$ *then* $\tau(f) = \tau(\mathscr{C})$ *where there is a basis-preserving[14] short exact sequence of* $Wh(G_L)$-*complexes*

$$0 \to C(\tilde{L}) \to \mathscr{C} \to \bar{C}(\tilde{K}) \to 0$$

PROOF: In computing $\tau(M_f, K)$ we may (19.3) use any universal covering of M_f. So, let us choose (see (3.14) and its proof) the natural projection $\alpha : M_{\tilde{f}} \to M_f$ such that $\alpha|\tilde{K}$ and $\alpha|\tilde{L}$ are the universal coverings of K and L implicit in our hypothesis. Let G be the group of covering homeomorphisms of $M_{\tilde{f}}$. If $\cdot g \in G_K$ define $E(g) \in G$ to be the unique extension of g to $M_{\tilde{f}}$. If $h \in G$ define $R(h) \in G_L$ to be restriction of h to $\tilde{L}$. It is an exercise [use (3.16), as in proving (20.2)] that in the commutative diagram

$$
\begin{array}{ccccc}
G_K & \xrightarrow{i_\#} & G & \xrightarrow{p_\#} & G_L \\
\Big\uparrow{\scriptstyle\theta(x,\tilde{x})} & & \Big\uparrow{\scriptstyle\theta(x,\tilde{x})} & & \Big\uparrow{\scriptstyle\theta(y,\tilde{y})} \\
\pi_1(K, x) & \xrightarrow{i_\#} & \pi_1(M_f, x) & \xrightarrow{p_\#} & \pi_1(L, y)
\end{array}
$$
$$\underbrace{\hspace{6cm}}_{f_\#}$$

we have $E = i_\# : G_K \to G$ and (because $\tilde{f}(\tilde{x}) = \tilde{y}$) $R = p_\# : G \to G_L$. Hence $f_\# = RE : G_K \to G_L$.

We view $C(M_{\tilde{f}}, \tilde{K})$ as a $Wh(G)$-complex, so that $E_*^{-1}(\tau C(M_{\tilde{f}}, \tilde{K}) = \tau(M_f, K) \in Wh(G_K)$ and $\tau(f) = f_*\tau(M_f, K) = R_*(\tau C(M_{\tilde{f}}, \tilde{K}))$. Thus by (18.2), $\tau(f) = \tau(\mathscr{C})$ where $\mathscr{C}$ is the $Wh(G_L)$-complex $[C(M_{\tilde{f}}, \tilde{K})]_R$. To show that $\mathscr{C}$ satisfies the conclusion of our theorem, we first study $C(M_{\tilde{f}}, \tilde{K})$.

---

[14] As in the last part of (17.2).

$C(M_{\tilde{f}}, \tilde{K})$ is (see (3.9)) naturally isomorphic as a complex over $\mathbb{Z}$ to the well-known "algebraic mapping cone" of $\tilde{f}_*$ which is given by

$$C_n = C_{n-1}(\tilde{K}) \oplus C_n(\tilde{L})$$
$$\partial_n(a) = -d_{n-1}(a) + \tilde{f}_*(a), \quad a \in C_{n-1}(\tilde{K})$$
$$\partial_n(b) = d'_n(b), \qquad b \in C_n(\tilde{L}).$$

A typical cell e^{n-1} of K gives rise, upon choosing a fixed lift, to an element $\langle \tilde{e}^{n-1} \rangle$ of $C_{n-1}(\tilde{K})$. [We will suppress the characteristic maps here to simplify the notation.] The image of $\langle \tilde{e}^{n-1} \rangle$ under the isomorphism of (3.9) is the element $\langle \tilde{e}^{n-1} \times (0, 1) \rangle$ of $C_n(M_{\tilde{f}}, \tilde{K})$. Suppose that, when $d\langle \tilde{e}^{n-1} \rangle$ is written as a linear combination in $\mathbb{Z}(G_K)$ we get

$$d\langle \tilde{e}^{n-1} \rangle = \sum_{i,j} n_{ij} g_i \langle \tilde{e}_j^{n-2} \rangle, \quad g_i \in G_K.$$

Then, over the ring $\mathbb{Z}$ we get

$$d\langle \tilde{e}^{n-1} \rangle = \sum_{i,j} n_{ij} \langle g_i \tilde{e}_j^{n-2} \rangle.$$

Applying the isomorphism of (3.9), the corresponding boundary in $C(M_{\tilde{f}}, \tilde{K})$ is

$$\begin{aligned} \partial \langle \tilde{e}^{n-1} \times (0, 1) \rangle &= -\Big(\sum_{i,j} n_{ij} \langle g_i \tilde{e}_j^{n-2} \times (0, 1) \rangle \Big) + \tilde{f}_* \langle \tilde{e}^{n-1} \rangle \\ &= -\Big(\sum_{i,j} n_{ij} E(g_i) \langle \tilde{e}_j^{n-2} \times (0, 1) \rangle \Big) + \tilde{f}_* \langle \tilde{e}^{n-1} \rangle. \end{aligned}$$

The last equation gives the boundary with $\mathbb{Z}(G)$-coefficients, and this equation holds because $E(g_i)|(M_{\tilde{f}} - \tilde{L}) = E(g_i)|(\tilde{K} \times [0, 1)) = g_i \times 1_{[0,1)}$.

In the same vein, if $\tilde{f}_* \langle \tilde{e}^{n-1} \rangle$ is written as a linear combination with coefficients in $\mathbb{Z}(G_L)$, and if the cells in L are denoted by u's, we get

$$\begin{aligned} \tilde{f}_* \langle \tilde{e}^{n-1} \rangle &= \sum_{p,q} m_{pq} \langle h_p \tilde{u}_q^{n-1} \rangle, \quad h_p \in G_L \\ &= \sum_{p,q} m_{pq} \langle (R^{-1} h_p) \tilde{u}_q^{n-1} \rangle, \quad \text{since } h_p | \tilde{L} = (R^{-1} h_p) | \tilde{L} \\ &= \sum_{p,q} m_{pq} (R^{-1} h_p) \langle \tilde{u}_q^{n-1} \rangle. \end{aligned}$$

A similar discussion holds for $\partial_n \langle \tilde{u}_n \rangle$ and we conclude that the matrix, when $C(M_{\tilde{f}}, \tilde{K})$ is considered as a $\mathbb{Z}(G)$-module, of $\partial_n : C_n(M_{\tilde{f}}, \tilde{K}) \to C_{n-1}(M_{\tilde{f}}, \tilde{K})$ is given by

$$\langle \partial_n \rangle = \left(\begin{array}{c|c} -E\langle d_{n-1} \rangle & R^{-1} \langle \tilde{f}_* \rangle \\ \hline \bigcirc & R^{-1} \langle d'_n \rangle \end{array} \right)$$

where $\langle d_{n-1} \rangle$ is a $\mathbb{Z}(G_K)$ matrix and $\langle \tilde{f}_* \rangle$ and $\langle d'_n \rangle$ are $\mathbb{Z}(G_L)$ matrices. It follows then from the equation $f_\# = RE$ that the complex $\mathscr{C} = C(M_{\tilde{f}}, \tilde{K})_R$, with $\tau(f) = \tau(\mathscr{C})$, has the boundary operator of, and is simply isomorphic to, the complex given in the statement of the theorem.

The assertion that the sequence $0 \to C(\tilde{L}) \to \mathscr{C} \to \bar{C}(\tilde{K}) \to 0$ is exact and basis-preserving follows immediately from the first part of the theorem. □

§23. Product and sum theorems

(23.1) (The sum theorem) *Suppose $K = K_1 \cup K_2$, $K_0 = K_1 \cap K_2$, $L = L_1 \cup L_2$, $L_0 = L_1 \cap L_2$ and that $f: K \to L$ is a map which restricts to homotopy equivalences $f_\alpha: K_\alpha \to L_\alpha (\alpha = 0, 1, 2)$. Let $j_\alpha: L_\alpha \to L$ and $i_\alpha: K_\alpha \to K$ be the inclusions. Then f is a homotopy equivalence and*

(a) $\tau(f) = j_{1*}\tau(f_1) + j_{2*}\tau(f_2) - j_{0*}\tau(f_0)$

(b) *If f is an inclusion map,*

$$\tau(L, K) = i_{1*}\tau(L_1, K_1) + i_{2*}\tau(L_2, K_2) - i_{0*}\tau(L_0, K_0)$$

PROOF: Let M_α denote the mapping cylinder of $f_\alpha (\alpha = 0, 1, 2)$. Then $M_\alpha \searrow K_\alpha$ since f_α is a homotopy equivalence. It follows that $(M_0 \cup K_1) \searrow K_1$. So, by the exact sequence of the triple $(M_1, M_0 \cup K_1, K_1)$, we have $\pi_i(M_1, M_0 \cup K_1) = 0$ for all i. Hence (3.2), $M_1 \searrow (M_0 \cup K_1)$. Similarly $M_2 \searrow (M_0 \cup K_2)$. Then $M_f = (M_1 \cup M_2) \searrow (M_1 \cup K) \searrow (M_0 \cup K) \searrow K$; whence f is a homotopy equivalence.

Now (20.2) and (20.4) give us:

(1) $\tau(M_f, K) = D_*\tau(M_f, M_0 \cup K) + \tau(M_0 \cup K, K)$

(2) $\tau(M_f, M_0 \cup K) = \tau(M_1 \cup K, M_0 \cup K) + \tau(M_2 \cup K, M_0 \cup K)$

(3) $\tau(M_\alpha \cup K, K) = D_*\tau(M_\alpha \cup K, M_0 \cup K) + \tau(M_0 \cup K, K)$, $\alpha = 1, 2$

where $D: M_0 \cup K \to K$ is a deformation retraction. Consequently

$$\tau(M_f, K) = \tau(M_1 \cup K, K) + \tau(M_2 \cup K, K) - \tau(M_0 \cup K, K)$$

Note that $fi_\alpha = j_\alpha f_\alpha$ and, using (20.3), that $\tau(M_\alpha \cup K, K) = i_{\alpha*}\tau(M_\alpha, K_\alpha)$. Thus, if we apply f_* to the last equation we get

$$\begin{aligned}\tau(f) &= (fi_1)_*\tau(M_1, K_1) + (fi_2)_*\tau(M_2, K_2) - (fi_0)_*\tau(M_0, K_0)\\ &= (j_1 f_1)_*\tau(M_1, K_1) + (j_2 f_2)_*\tau(M_2, K_2) - (j_0 f_0)_*\tau(M_0, K_0)\\ &= j_{1*}\tau(f_1) + j_{2*}\tau(f_2) - j_{0*}\tau(f_0), \text{ proving (a).}\end{aligned}$$

Finally, assertion (b) follows from (a) and (22.3) and the observation that $f_*^{-1} j_{\alpha*} f_{\alpha*} = i_{\alpha*}$. □

The behavior of torsion under the taking of Cartesian products is quite interesting. For example, if $K \searrow K_0$ then, regardless of what $\tau(K, K_0)$ is, we have $\tau(K \times S^1, K_0 \times S^1) = 0$ where S^1 is the 1-sphere. The complete picture is given by[15]

(23.2) (The product theorem) (a) *If P, K, K_0 are finite CW complexes where $K \searrow K_0$ and P is connected then*

$$\tau(K \times P, K_0 \times P) = \chi(P) \cdot i_* \tau(K, K_0)$$

where $i: K_0 \to K_0 \times P$ by $i(x) = (x, y)$ for some fixed y, and $\chi(P)$ denotes the Euler characteristic of P.

(b) *If $f \times g: K \times K' \to L \times L'$ where f, g are homotopy equivalences between connected complexes and if $i: L \to L \times L'$ and $j: L' \to L \times L'$ as in (a) then*

$$\tau(f \times g) = \chi(L') \cdot i_* \tau(f) + \chi(L) \cdot j_* \tau(g)$$

PROOF[16] *OF* (a)*:* We start with two preliminary remarks:

First if P is not connected but instead has components $P_1, P_2, \ldots, P_q$, the connected case immediately implies that

$$\tau(K \times P, K_0 \times P) = \sum_j \chi(P_j) \cdot i_{j*} \tau(K, K_0)$$

where $i_j: K_0 \to K_0 \times P_j$ by $i_j(x) = (x, y_j)$ for fixed y_j.

Secondly, if the assertion (a) is true for a complex Q simple-homotopy equivalent to P, it is true for P. For suppose that $f: Q \to P$ is a cellular simple-homotopy equivalence. Then (5.8) $M_f \searrow Q$ rel Q. Hence (exercise) $K \times M_f \searrow K \times Q$ rel $K \times Q$. But $K \times M_f = M_{1_K \times f}$, so $1_K \times f: K \times Q \to K \times P$ is a simple-homotopy equivalence. Similarly $1_{K_0} \times f$ is a simple-homotopy equivalence. Denote these by F and $\bar{F}$ respectively. By assumption $\tau(K \times Q, K_0 \times Q) = \chi(Q) \cdot \hat{\imath}_* \tau(K, K_0)$ where $\hat{\imath}(x) = (x, y)$ for some fixed y; and by (22.7), $\tau(K \times P), K_0 \times P) = \bar{F}_* \tau(K \times Q, K_0 \times Q)$. Hence $\tau(K \times P, K_0 \times P) = \chi(Q) \cdot (\bar{F}\hat{\imath})_* \tau(K, K_0) = \chi(P) \cdot i_* \tau(K, K_0)$ where $i(x) = (x, f(y))$ for all x.

The proof of (a) now proceeds by induction on $n = \dim P$. When $n = 0$, P is a point and the result is trivial. Suppose $n > 0$ and the result is known for integers less than n. Let $e_1, \ldots, e_q$ be the n-cells of P, with characteristic maps $\varphi_\alpha: I^n \to P$. Fixing a CW structure on ∂I^n, we may assume that the attaching maps $\varphi_\alpha | \partial I^n$ are cellular. For, if not, we could homotop each of them to a cellular map to get, by (7.1), a new complex simple-homotopy equivalent to P and then prove the assertion for this new complex. Taking q disjoint copies $\{I^n_\alpha\}$ of I^n, define $\varphi: (\partial I^n_1) \oplus \ldots \oplus (\partial I^n_q) \to P^{n-1}$ by the condition $\varphi | \partial I^n_\alpha = \varphi_\alpha | \partial I^n_\alpha$. Then M_φ is a CW complex and we set $Q = M_\varphi \cup I^n_1 \cup \ldots \cup I^n_q$ where I^n_α is attached to M_φ by the identity along ∂I^n_α. Q is simple-homotopy equivalent to P by the simple-homotopy extension theorem (5.9), since the natural projection $p: M_\varphi \to P^{n-1}$ is a cellular simple-homotopy equivalence and $Q \cup_\varphi P^{n-1}$ is isomorphic to P. Thus we may prove our assertion for Q.

---

15 For the generalization of this to fiber bundles see [ANDERSON 1, 2, 3].

16 The idea here is due to D. R. Anderson. Other proofs have been given in [KWUN-SZCZARBA], [STÖCKER] and [HOSOKAWA].

Let $R = I_1^n \cup \ldots \cup I_q^n$, a subcomplex of Q, and let $\partial R = \partial I_1^n \cup \ldots \cup \partial I_q^n$. So $Q = M_\varphi \cup R$ and $M_\varphi \cap R = \partial R$. If $n > 1$ choose constant sections $j: K_0 \to K_0 \times M_\varphi$, $k_\alpha: K_0 \to K_0 \times \partial I_\alpha^n$ and $m_\alpha: K_0 \to K_0 \times I_\alpha^n$ $(1 \leq \alpha \leq q)$. If $n = 1$, notice that M_φ has as many components $M_{\varphi,\beta}$ as P^0 has points—say r components—and let $j_\beta: K_0 \to K_0 \times M_{\varphi,\beta}$ be constant sections $(1 \leq \beta \leq r)$. Let $k_{\alpha,1}$ and $k_{\alpha,2}$ be constant sections into the components of $K_0 \times \partial I_\alpha^1$ and let $m_\alpha: K_0 \to K_0 \times I_\alpha^1$ also be constant sections $(1 \leq \alpha \leq q)$.

First consider the case $n \neq 1$. From the sum theorem we have

$$(1)\ \tau(K \times Q, K_0 \times Q) = f_{1*}\tau(K \times M_\varphi, K_0 \times M_\varphi) + f_{2*}\tau(K \times R, K_0 \times R) - f_{0*}\tau(K \times \partial R, K_0 \times \partial R)$$

where f_0, f_1, f_2 are inclusion maps into $K_0 \times Q$. But

$$(2)\ \tau(K \times M_\varphi, K_0 \times M_\varphi) = \chi(M_\varphi) \cdot j_*\tau(K, K_0)$$

$$(3)\ \tau(K \times R, K_0 \times R) = \sum_\alpha m_{\alpha*}\tau(K, K_0)$$

$$(4)\ \tau(K \times \partial R, K_0 \times \partial R) = \sum_\alpha \chi(S^{n-1}) \cdot k_{\alpha*}\tau(K, K_0).$$

(2) holds because M_φ has the same simple-homotopy type as P^{n-1}, to which the induction hypothesis applies. (3) comes from the first preliminary remark and the fact that each component of R has the simple-homotopy type of a point. (4) follows by induction because each component of ∂R is an $(n-1)$-sphere. But now the connectedness of P implies that all the maps $f_1 j, f_2 m_\alpha$, and $f_0 k_\alpha$ are homotopic to any given constant section $i: K_0 \to K_0 \times P$. Hence, substituting into (1),

$$\begin{aligned}\tau(K \times Q, K_0 \times Q) &= [\chi(M_\varphi) + q - q\chi(S^{n-1})] i_*\tau(K, K_0) \\ &= [\chi(M_\varphi) + q(-1)^n] i_*\tau(K, K_0) \\ &= \chi(Q) \cdot i_*\tau(K, K_0).\end{aligned}$$

In the case $n = 1$ the equations above become

$$(2')\ \tau(K \times M_\varphi, K_0 \times M_\varphi) = \textstyle\sum_{\beta=1}^r j_{\beta*}\tau(K, K_0)$$

$$(3')\ \tau(K \times R, K_0 \times R) = \textstyle\sum_{\alpha=1}^q m_{\alpha*}\tau(K, K_0)$$

$$(4')\ \tau(K \times \partial R, K_0 \times \partial R) = \textstyle\sum_{\alpha=1}^q [k_{\alpha,1}\tau(K, K_0) + k_{\alpha,2}\tau(K, K_0)]$$

Using the connectedness of P as above these yield

$$\begin{aligned}\tau(K \times Q, K_0 \times Q) &= (r + q - 2q) \cdot i_*\tau(K, K_0) \\ &= \chi(P) \cdot i_*\tau(K, K_0) \\ &= \chi(Q) \cdot i_*\tau(K, K_0).\end{aligned}$$

PROOF OF (b): We must find $\tau(f \times g)$ where $(f \times g): K \times K' \to L \times L'$. First consider the special case $K' = L' = P$, $g = 1_P$. Then

$$\begin{aligned}\tau(f\times 1_P) &= (f\times 1_P)_*\tau(M_{f\times 1_P}, K\times P)\\ &= (f\times 1_P)_*\tau(M_f\times P, K\times P)\\ &\overset{(a)}{=} \chi(P)\cdot(f\times 1_P)_*\alpha_*\tau(M_f, K) \text{ where } \alpha: K\to K\times P \text{ by } \alpha(x) = (x, p_0)\\ &= \chi(P)\cdot(i\circ f)_*\tau(M_f, K) \text{ since } (i\circ f) = (f\times 1_P)\circ\alpha\\ &= \chi(P)\cdot i_*\tau(f).\end{aligned}$$

The general case now follows easily from the fact that

$$(f\times g) = (1_L\times g)\circ(f\times 1_{K'}). \quad \square$$

§24. The relationship between homotopy and simple-homotopy

We first show that any torsion element can be realized as the torsion of some homotopy equivalence. Thus Conjecture **I** of §4 ("Every homotopy equivalence is a simple-homotopy equivalence") is decidedly false.

(24.1) *If $\tau_0 \in Wh(L)$ then there is a CW complex K and a homotopy equivalence $f: K\to L$ with $\tau(f) = \tau_0$.*

PROOF: Let K be a CW complex such that $K \searrow L$ and such that $\tau(K, L) = -\tau_0$. Such a complex exists by the first definition (§6) of $Wh(L)$. Let f: $K\to L$ be a homotopy inverse to the inclusion map $i: L\to K$. Then (22.3)–(22.5) yield $\tau(f) = -f_*\tau(i) = -f_*i_*\tau(K, L) = -\tau(K, L) = \tau_0$. $\square$

Conjecture **II** of §4 ("If there exists a homotopy equivalence $f: K\to L$ then there exists a simple-homotopy equivalence") is more elusive. Its answer depends not only on $Wh(L)$, but also on how rich is the group $\mathscr{E}(L)$ of equivalence classes (under homotopy) of self-homotopy equivalences of L. This is explained by the next three propositions.[17]

(24.2) *Suppose that L is a given CW complex. If K is homotopy equivalent to L (written "$K \simeq L$") define $S_K \subset Wh(L)$ by*

$$S_K = \{\tau(f) \mid f: K\to L \text{ is a homotopy equivalence}\}.$$

Then, if $K \simeq L \simeq K'$, the following are equivalent assertions:

(a) $S_K \cap S_{K'} \neq \emptyset$.

(b) *K and K' have the same simple-homotopy type.*

(c) $S_K = S_{K'}$.

Thus $\mathscr{F} = \{S_K \mid K \simeq L\}$ is a family of sets which partitions $Wh(L)$. The cardinality of $\mathscr{F}$ is exactly that of the set of simple-homotopy equivalence classes within the homotopy equivalence class of L.

---

[17] Compare [Cockroft-Moss].

PROOF: (a) $\Rightarrow$ (b). Suppose that $S_K \cap S_{K'} \neq \varnothing$. Then there are homotopy equivalences $f: K \to L$ and $g: K' \to L$ such that $\tau(f) = \tau(g)$. Let $\bar{g}$ be a homotopy inverse to g. Then $\bar{g}f: K \to K'$, and by (22.4) and (22.5),

$$\tau(\bar{g}f) = \tau(\bar{g}) + \bar{g}_*\tau(f) = -\bar{g}_*\tau(g) + \bar{g}_*\tau(f) = 0.$$

(b) $\Rightarrow$ (c): Suppose that $s: K' \to K$ is a simple homotopy equivalence. If $\tau_0 \in S_K$ choose $f: K \to L$ with $\tau(f) = \tau_0$. Then $fs: K' \to L$ and $\tau(fs) = \tau(f) + f_*\tau(s) = \tau(f) = \tau_0$. Thus $S_K \subset S_{K'}$. By symmetry $S_K = S_{K'}$.
(c) $\Rightarrow$ (a): This is trivial since, by definition, $S_K \neq \varnothing$. □

Exercise: ([Cockroft-Moss]) The sets S_K are the orbits of the action of $\mathscr{E}(L)$ on $Wh(L)$ given by $f \cdot \alpha = \tau(f) + f_*(\alpha)$.†

Let us adopt the notation:

$|S|$ = cardinality of the set S

$\nu_L = |\mathscr{F}|$, $\mathscr{F}$ as in (24.2)

$\mathscr{E}(L)$ = the group of equivalence classes (under homotopy) of self-homotopy equivalences of L

$Wh_0(L) = \{\tau(f) \in Wh(L) \mid f_*: Wh(L) \to Wh(L) \text{ is the identity}\}$.

Notice that $Wh_0(L)$ is a subgroup of $Wh(L)$.

(24.3) $\nu_L \cdot |Wh_0(L)| \leq |Wh(L)| \leq \nu_L \cdot |\mathscr{E}(L)|$.

PROOF: If $g: K \to L$ is a fixed homotopy equivalence then the correspondence $f \to fg$ (f a self homotopy equivalence of L) induces a bijection of $\mathscr{E}(L)$ to the set $\mathscr{E}(K, L)$ of equivalence classes of homotopy equivalences $K \to L$. Thus, by (22.1), $|S_K| \leq |\mathscr{E}(K, L)| = |\mathscr{E}(L)|$ for all K, and the inequality $|Wh(L)| \leq \nu_L \cdot |\mathscr{E}(L)|$ follows from (24.2).

On the other hand, if $g_0: K \xrightarrow{\simeq} L$ then, for any f which induces the identity on $Wh(L)$, we have $\tau(fg_0) = \tau(f) + \tau(g_0) \in S_K$. So the coset $\tau(g_0) + Wh_0(L)$ is contained in S_K, and $|S_K| \geq |Wh_0(L)|$. Hence, from (24.2), $\nu_L \cdot |Wh_0(L)| \leq |Wh(L)|$. □

(24.4) *Suppose that L is a CW complex. Then*

(1) *[Every complex homotopy equivalent to L is simple-homotopy equivalent to L]* $\Leftrightarrow$ $[Wh(L) = \{\tau(f) \mid f \in \mathscr{E}(L)\}]$.

(2) *If $Wh(L)$ is infinite and $\mathscr{E}(L)$ is finite, there are infinitely many simple-homotopy equivalence classes within the homotopy equivalence class of L.*

(3) *Every finite connected 2-complex L with $\pi_1 L \cong \mathbb{Z}_p$, $p \neq 1, 2, 3, 4, 6$ is a space with infinite Whitehead group satisfying the conditions of* (1). *Every lens space $L = L(p; q_1, q_2, \ldots, q_n)$, $p \neq 1, 2, 3, 4, 6$, satisfies the hypothesis of* (2).

PROOF: The assertions in (1) are logically equivalent because, by (24.2), each is equivalent to the assertion that $S_K = S_L$ for all K.

---

† *Added in Proof:* P. Olum has shown, when $L = L_{5,2}$ (§27), that $|S_L| = 1$ while $|S_K| = 2$ if $S_K \neq S_L$. This implies that $|\{\tau(f) \mid f \in \mathscr{E}(L)\}| \neq |\{\tau(f) \mid f \in \mathscr{E}(K)\}|$ although $K \simeq L$!

(2) follows trivially from (24.2) and (24.3).

To prove (3), note that if p is a positive integer, $p \neq 1, 2, 3, 4, 6$ and if $\mathbb{Z}_p$ is the cyclic group of order p then $Wh(\mathbb{Z}_p)$ is infinite. (See (11.4) and (11.5).) In Chapter V, on the other hand, we shall discuss the lens spaces $L = L(p; q_1, \ldots, q_n)$ and show that $\pi_1 L = \mathbb{Z}_p$, and that $\mathscr{E}(L)$ is in one-one correspondence with $\{a | 0 < a < p, a^n \equiv \pm 1 (\text{mod } p)\}$. Thus the hypothesis of (2) is satisfied by these lens spaces.

The pseudo-projective plane P_p is the 2-dimensional complex gotten by attaching a single 2-cell to the unit circle S^1 by the map $f: S^1 \to S^1$ which is given in complex coordinates by $f(z) = z^p$. Clearly $\pi_1(P_p) = \mathbb{Z}_p$. PAUL OLUM studies the pseudo-projective planes in [OLUM 1, 2] and shows that any $\tau_0 \in Wh(P_p)$ is the torsion of a self equivalence $f: (P_p, e^0) \to (P_p, e^0)$ such that f induces the identity on $\pi_1 (P_p, e^0)$. Thus these spaces satisfy the assertions of (1).

Finally, [DYER-SIERADSKI] shows that any finite connected 2-complex with finite cyclic fundamental group $\mathbb{Z}_p$ is homotopy equivalent to a complex of the form $P_p \vee S^2 \vee S^2 \vee \ldots \vee S^2$. Thus, as these authors point out (and the reader should verify), OLUM's work implies that the assertions of (1) are satisfied by any such 2-complex. ([DYER-SIERADSKI] also proves this directly.) □

At present it is unknown whether homotopy type equals simple-homotopy type for *arbitrary* finite 2-complexes.

§25. Invariance of torsion, h-cobordisms and the Hauptvermutung

The following question is still unanswered in general.†

Topological invariance of Whitehead torsion: *If $h: K \to L$ is a homeomorphism between finite CW complexes, does it follow that $\tau(h) = 0$?*

In this section we shall give affirmative answers in some very special cases and try to indicate how this relates to some of the most exciting developments in modern topology.

Definition: A subdivision of the CW complex K is a pair (K_*, h) where K_* is a CW complex and $h: K_* \to K$ is a homeomorphism such that for each cell e of K_* there is a cell e' of K with $h(e) \subset e'$. (As always, "cell" means "open cell".)

(25.1) *If (K_*, h) is a subdivision of K then h is a simple-homotopy equivalence.*

PROOF: Let $g = h^{-1}: K \to K_*$. Clearly g is a cellular map and it suffices, by (22.5), to show that $\tau(g) = 0$ or, what is the same thing, that $\tau(M_g, K) = 0$.

Let $K = e_0 \cup e_1 \cup \ldots \cup e_n = K_n$, where $\dim e_j \leq \dim e_{j+1}$. Let $K_j = e_0 \cup \ldots \cup e_j$ and let M_j be the mapping cylinder of the induced map $K_j \to g(K_j)$. Then (M_j, K_j) is homeomorphic to $(K_j \times I, K_j \times 0)$, so $M_j - M_{j-1} \approx e_j \times (0,1]$ and (20.1) and (20.2) imply that

$$\begin{aligned}\tau(M_j \cup K, K) &= \tau(M_{j-1} \cup K, K) + i_*^{-1}\, \tau(M_j \cup K, M_{j-1} \cup K) \\ &= \tau(M_{j-1} \cup K, K).\end{aligned}$$

† *Added in proof:* An affirmative response, due to T. Chapman, is given in the Appendix.

Thus, starting with $(M_g, K) = (M_n, K)$, an induction argument shows that $\tau(M_g, K) = \tau(K, K) = 0$. □

(25.2) *If* $\tau(K, L) = 0$ *and if* (K_*, L_*) *subdivides* (K, L) [i.e. there is *a subdivision* (K_*, h) *of* K *with* $L_* = h^{-1}(L)$] *then* $\tau(K_*, L_*) = 0$.

PROOF: Let $\bar{h}: L_* \to L$ be the restriction of h. Clearly $(L_*, \bar{h})$ is a subdivision of L, so $\bar{h}$ is a simple-homotopy equivalence. Hence, by (22.7), $\tau(K_*, L_*) = \bar{h}_*^{-1}\tau(K, L) = 0$. □

The invariance of torsion under subdivision is of importance in piecewise-linear (and, consequently, in differential) topology. If K and L are finite simplicial complexes then a map: $f: |K| \to |L|$ is *piecewise linear* (p. 1.) if there are simplicial subdivisions K_* and L_* $\{(K_*, 1_{|K|})$ and $(L_*, 1_{|L|})$ in the notation of the preceding paragraphs$\}$ such that $f: K_* \to L_*$ is a simplicial map.[18] Our results on CW subdivision easily imply

(25.3) *If* $h: K \to L$ *is a p.l. homeomorphism then* $\tau(h) = 0$. *If* $h: (K, K_0) \to (L, L_0)$ *is a p.l. homeomorphism of pairs then* $\tau(K, K_0) = 0$ *if and only if* $\tau(L, L_0) = 0$. □

Recent results, which we cannot prove here, show that the assumption that h is *p.l.* can sometimes be dropped. These results are summarized by

(25.4): *Suppose that* $h: K \to L$ *is a homeomorphism between polyhedra. If either* (a) *dim* $K =$ *dim* $L \leq 3$ *or* (b) K *and* L *are p.l. manifolds*[19] *then* $\tau(h) = 0$.

REFERENCES: (a) follows from the result of [Brown] that every homeomorphism between polyhedra of dimension ≤ 3 is isotopic to a *p.l.* homeomorphism. (b) is a result of [Kirby-Siebenmann] (despite the fact that they also have examples of *p.l.* manifolds which are homeomorphic but not *p.l.* homeomorphic!) □

The Kirby-Siebenmann examples mentioned in the last paragraph are counterexamples to the following classical conjecture:

> **The Hauptvermutung:** *If* P *and* Q *are homeomorphic finite simplicial complexes then they are p.l. homeomorphic.*

The first counterexample to this conjecture was given in [Milnor 2]. Later Stallings showed ([Stallings 2]) how Milnor's idea could be used to generate myriads of examples. Proceeding in their spirit we now explain how torsion, and in particular (25.3), can be used to construct counterexamples to the Hauptvermutung. Crucial to this approach and fundamental in the topology of manifolds is the concept of an h-cobordism.

An *h-cobordism* is a triple (W, M_0, M_1) where W is a compact *p.l.* $(n+1)$-manifold whose boundary consists of two components, M_0 and M_1

---

[18] See [Hudson] for an exposition of the piecewise-linear category.
[19] For a definition see [Hudson] or §26.

with $W \searrow M_0$ and $W \searrow M_1$. The following are basic facts about h-cobordisms. (In each case we assume $n = \dim M_0$.)

(A) *If* $n \geq 4$ *it follows* (from "engulfing") *that* $W - M_0 \overset{p.1.}{\cong} M_1 \times (0, 1]$ *and* $W - M_1 \overset{p.1.}{\cong} M_0 \times [0, 1)$, *and from this that*

$$a{*}M_0 \cup b{*}M_0 = \text{susp}\,(M_0) \overset{\text{top}}{\cong} c{*}M_0 \cup W \cup d{*}M_1$$

where a, b, c, d *are points, "susp" denotes suspension and "$*$" means "join".*

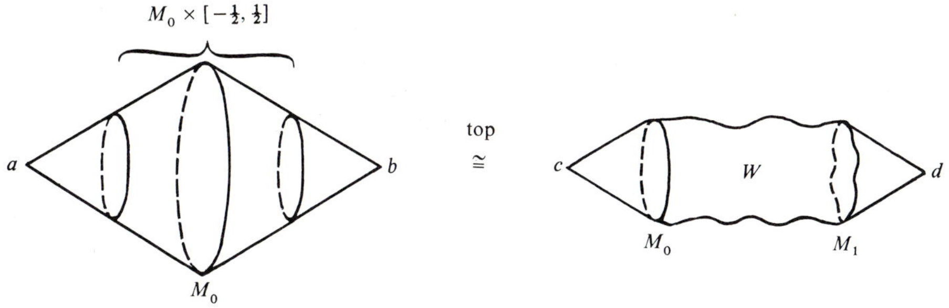

{A reference is [Hudson; Part 1, Theorem 7.11].}

(B) The s-cobordism theorem: *If* $n \geq 5$ *and* $\tau(W, M_0) = 0$ *then* $(W, M_0, M_1) \overset{p.1.}{\cong} (M_0 \times I, M_0 \times 0, M_0 \times 1)$ {The proof is analogous to the proof in §7 and §8 that if $\tau(K, L) = 0$ then $K \searrow L$ rel L. One trades and cancels handles rather than cells. Reference: [Hudson; Part 2, Theorem 10.10].}

(C) Realization: *If* M_0 *is a closed p.*1. *n-manifold, where* $n \geq 5$, *and if* $\tau_0 \in Wh(M_0)$ *then there is an h-cobordism* (W, M_0, M_1) *with* $\tau(W, M_0) = \tau_0$. {The proof is analogous to that of (8.7). Reference: [Hudson; Part 2, Theorem 12.1].}

(D) Classification: *If* (W, M_0, M_1) *and* (W', M_0, M_1') *are h-cobordisms with* $\tau(W, M_0) = \tau(W', M_0)$ *and if* $n \geq 5$ *then* $(W, M_0, M_1) \overset{p.l.}{\cong} (W', M_0, M_1')$. {This follows from (*B*) and (*C*). Reference: [Hudson; Part 2, Theorem 12.2]}.

(E) Duality: *If* (W, M_0, M_1) *is an h-cobordism then* $i_{0*}\tau(W, M_0) = (-1)^n$ $i_{1*}\tau^*(W, M_1)$ *where* $i_j : M_j \xrightarrow{\subset} W$, $j = 0, 1$, *and* $\tau^*(W, M_1)$ *is the image of* $\tau(W, M_1)$ *under "conjugation" of* $Wh(M_1)$, *or "twisted conjugation" if* M_1 *is not orientable.* {Reference: Hudson; page 273}.

Now, using (*C*) let (W, M_0, M_1) be an h-cobordism with $\tau(W, M_0) \neq 0$, and let $V = c{*}M_0 \cup W \cup d{*}M_1$. By (A), susp (M_0) is homeomorphic to V. Suppose that there were a $p.$1. homeomorphism h:susp $(M_0) \to V$. Then $h(\{a, b\}) = \{c, d\}$ since these are the points where the spaces fail to be topological manifolds. Thus, being $p.$1., h would take a regular neighborhood (see [Cohen]) of $\{a, b\}$ to a regular neighborhood of $\{c, d\}$ and, from the equivalence of regular neighborhoods via ambient $p.1.$ homeomorphism, we could assume that h restricts to a $p.1.$ homeomorphism

$$(M_0 \times [-\tfrac{1}{2}, \tfrac{1}{2}], M_0 \times \{-\tfrac{1}{2}\}) \overset{p.1.}{\cong} (W, M_0).$$

Then (25.3) implies that $\tau(W, M_0) = 0$, which contradicts the choice of W. Thus V and susp (M_0) are not $p.1.$ homeomorphic, although they are homeomorphic.

It is interesting to note [MILNOR 1; p. 400] that there are h-cobordisms (W, M_0, M_1) with $M_0 \overset{p.1.}{\cong} M_1$ and $\tau(W, M_0) \neq 0$. Using such an h-cobordism in the preceding construction, one can conclude that susp (M_0) and V are examples of spaces which are homeomorphic and locally $p.1.$ homeomorphic, but are not $p.1.$ homeomorphic. Of course the later KIRBY-SIEBENMANN examples, which are $p.1.$ manifolds, are more striking illustrations of this phenomenon.

Chapter V

Lens Spaces

§26. Definition of lens spaces

In this chapter we give a detailed introduction to the theory of lens spaces.[20] These spaces are fascinating in their own right and will supply examples on which to make the preceding theory concrete.

We shall at times use the language and setting of the piecewise linear ($p.$l.) category. (See [HUDSON]). However, the reader who is willing to settle for "manifolds" and "maps" whenever "$p.$l. manifolds" and "$p.$l. maps" appear, can proceed with equanimity.

A $p.$l. *n-manifold* (without boundary) is a pair $(M, \mathscr{A})$ where M is a separable metric space and $\mathscr{A}$ is a family of pairs (U_i, h_i) such that $\{U_i\}$ is an open cover of M, $h_i: U_i \to R^n$ is a homeomorphism onto an open subset of R^n, and $h_j h_i^{-1}: h_i(U_i \cap U_j) \to R^n$ is $p.$l. for all i, j. $\mathscr{A}$ is called a $p.$l. *atlas* and the (U_i, h_i) are called *coordinate charts*.

If M_1 and M_2 are $p.$l. manifolds of dimensions m and n respectively then $f: M_1 \to M_2$ is a $p.$l. *map* if, for each $x \in M_1$, there is a coordinate chart (U, h) about x and a coordinate chart (V, g) about $f(x)$ such that the map $gfh^{-1}: h[U \cap f^{-1}V] \to R^m$ is $p.$l.

If M is a topological space and G is a group of auto-homeomorphisms of M, then G *acts freely on* M if: $[x \in M,\ 1 \neq g \in G] \Rightarrow [g(x) \neq x]$. The set $G(x_0) = \{g(x_0) | g \in G\}$ is called the *orbit* of x_0 under G. We denote by M/G the quotient space of M under the equivalence relation: $x \sim y \Leftrightarrow G(x) = G(y)$. Thus the points of M/G are the orbits under G.

(26.1) *If M is a connected* $p.$l. *manifold and G is a finite group of* $p.$l. *homeomorphisms acting freely on M then*

(a) *The quotient map $\pi: M \to M/G$ is a covering map.*

(b) *The group G is precisely the group of covering homeomorphisms.*

(c) *π induces a* $p.$l. *structure on M/G with respect to which π is* $p.$l.

PROOF: (a) and (b) are left as exercises for the reader (or see [SPANIER, p. 87]). To prove (c), let $\{(U_i, h_i)\}_{i \in J}$ be a $p.$l. atlas for M with coordinate charts chosen small enough that $\pi | U_i: U_i \to \pi(U_i)$ is a homeomorphism for each U_i. Denote $\pi_i = \pi | U_i$. Then $\{(\pi(U_i), h_i \pi_i^{-1})\}_{i \in J}$ is a $p.$l. atlas for M/G. To prove this we must show that, for $i, j \in J$, the homeomorphism $h_j \pi_j^{-1} \pi_i h_i^{-1}$, with domain $h_i \pi_i^{-1}[\pi(U_i) \cap \pi(U_j)]$, is $p.$l. But on each component of

[20] A more advanced treatment which goes much further is given in [MILNOR 1].

$\pi_i^{-1}[\pi(U_i) \cap \pi(U_j)]$ the homeomorphism $\pi_j^{-1}\pi_i$ agrees with some element of G. Since the elements of G are *p.l.*, $\pi_j^{-1}\pi_i$ is *p.l.* But h_j^{-1} and h_i are certainly *p.l.* So $h_j\pi_j^{-1}\pi_i h_i^{-1}$ is also *p.l.*, as desired.

We leave the reader to check that now $\pi: M \to M/G$ is *p.l.* □

Suppose that $p \geq 2$ is an integer (*not* necessarily prime) and that $q_1, q_2, \ldots, q_n$ are integers relatively prime to p. [i.e. $(p, q_j) = 1$ where (,) denotes the greatest common divisor.] Then the *lens space* $L(p; q_1, q_2, \ldots, q_n)$ is a $(2n-1)$-dimensional *p.l.* manifold which we now define as Σ^{2n-1}/G for appropriate Σ^{2n-1} and G.

If $p > 2$, let Σ^1 be the regular polygon (simplicial 1-sphere) in R^2 with vertices $v_q = e^{2\pi i q/p}$, $q = 0, 1, 2, \ldots, p-1$. Let Σ^{2n-1} be the polyhedron in $R^{2n} = R^2 \times R^2 \times \ldots \times R^2$ gotten by taking the iterated join

$$\begin{aligned}\Sigma^{2n-1} &= \Sigma_1 * \Sigma_2 * \ldots * \Sigma_n \\ &= \{\lambda_1 z_1 + \ldots + \lambda_n z_n \mid \sum_j \lambda_j = 1, \lambda_j \geq 0, z_j \in \Sigma_j\}\end{aligned}$$

Here Σ_j is the copy of Σ^1 in $\underbrace{0 \times 0 \times \ldots R^2}_{j} \times 0 \times \ldots \times 0$ and each $z \in \Sigma^{2n-1}$ is uniquely expressible as such a sum. Σ^{2n-1} is a simplicial complex and, as a join of circles, is a *p.l.* $(2n-1)$-sphere.

When $p = 2$ we must vary the above procedure (since two points don't determine a circle). Let Σ^1 be the regular polygon with vertices $v_0 = 1$, $A = e^{\pi i/2}$, $v_1 = e^{\pi i}$ and $B = e^{3\pi i/2}$. Σ^{2n-1} is then described as above.

To construct a group G which acts on Σ^{2n-1}, let R_j be the rotation of Σ_j by q_j notches, a *notch* consisting of $\dfrac{2\pi}{p}$ radians. Let $g = R_1 * R_2 * \ldots * R_n$: $\Sigma^{2n-1} \to \Sigma^{2n-1}$; i.e.,

$$g(\sum_j \lambda_j z_j) = \sum_j \lambda_j R_j(z_j)$$

As a join of simplicial isomorphisms g is a simplicial isomorphism. Clearly $g^p = 1$. But, if $1 \leq k \leq p-1$, g^k can fix no point of Σ^{2n-1}. For let $z = \sum_{j=1}^{n} \lambda_j z_j$ where $\lambda_{j_o} \neq 0$. Then

$$\begin{aligned}(q_{j_0}, p) = 1 &\Rightarrow (R_{j_0})^k(z_{j_0}) \neq z_{j_0} \\ &\Rightarrow g^k(z) = \sum_j \lambda_j R_j^k(z_j) \neq \sum_j \lambda_j z_j = z\end{aligned}$$

Hence $G = \{1, g, g^2, , \ldots g^{p-1}\}$ is a group of order p of *p.l.* homeomorphisms which acts freely on Σ^{2n-1}, and it is with this G that we define

$$L(p; q_1, \ldots, q_n) = \Sigma^{2n-1}/G$$

By (26.1), $L = L(p; q_1, \ldots, q_n)$ is a *p.l.* manifold and $\pi: \Sigma^{2n-1} \to L$ is a *p.l.* covering map with G as the group of covering transformations.

REMARK: $L(p; q_1, q_2, \ldots, q_n)$ can also be naturally defined as a differentiable manifold by thinking of it as a quotient of the (round) sphere S^{2n-1}. Let $\bar{g}$: $R^{2n} \to R^{2n}$ by $\bar{g}(z_1, z_2, \ldots, z_n) = (R_1(z_1), R_2(z_2), \ldots, R_n(z_n))$ where R_j is the rotation of R^2 through $(2\pi q_j/p)$ radians. Then $\bar{g}$ is an orthogonal transformation such that $\bar{g}, \bar{g}^2, \ldots, \bar{g}^{p-1}$, have no fixed points other than 0. Hence $\bar{G} = \{\bar{g}^k | S^{2n-1} : 0 \leq k \leq p-1\}$ is a group of diffeomorphisms of S^{2n-1} and $S^{2n-1}/\bar{G}$ is a smooth manifold (by a proof analogous to that of (26.1)).

The connection with Σ^{2n-1}/G is gotten by noting that $g = \bar{g}|\Sigma^{2n-1}$ and that, if $T: \Sigma^{2n-1} \to S^{2n-1}$ by $T(z) = z/|z|$, the following diagram commutes.

$$\begin{array}{ccc} \Sigma^{2n-1} & \xrightarrow[\approx]{T} & S^{2n-1} \\ {\scriptstyle g}\downarrow & & \downarrow{\scriptstyle \bar{g}|S^{2n-1}} \\ \Sigma^{2n-1} & \xrightarrow[\approx]{T} & S^{2n-1} \end{array}$$

From this it follows that there is a piecewise-differentiable homeomorphism $H: \Sigma^{2n-1}/G \to S^{2n-1}/\bar{G}$ which is covered by T. This homeomorphism can be used (as in the proof of (26.1c)) to give $S^{2n-1}/\bar{G}$ a $p.1.$ structure which is "compatible with its smooth structure" and with respect to which H is a $p.1.$ homeomorphism.

§27. The 3-dimensional spaces $L_{p,q}$

Let B^3 be the closed unit ball in R^3 and let D^2_+ and D^2_- be the closed upper and lower hemispheres of Bdy B^3. Suppose that integers p, q are given with $p \geq 2, (p, q) = 1$. Let R be the rotation of R^2 through $2\pi q/p$ radians, and define $h: D^2_- \to D^2_+$ by $h(x, y, z) = (R(x, y), -z)$. In this setting the 3-dimensional lens space $L_{p,q}$ is often defined (see [Seifert-Threllfall]) as the quotient space under the equivalence relation generated by h.

$$L_{p,q} = \frac{B^3}{[x \sim y \text{ if } x \in D^2_- \text{ and } y = h(x)]}$$

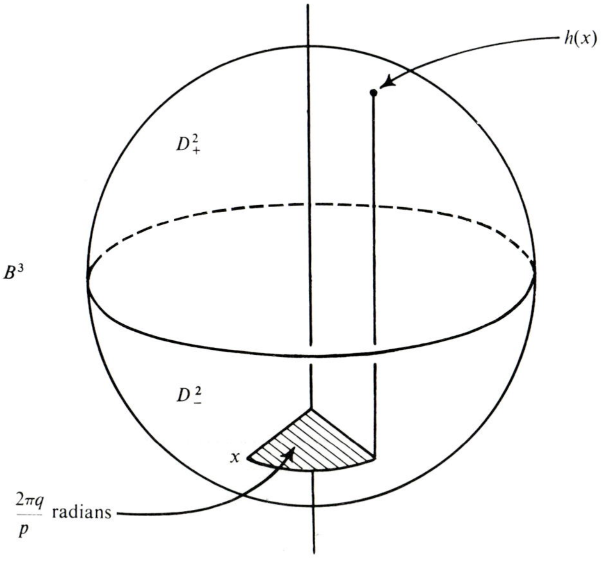

In this section we wish to point out intuitively why $L_{p,q}$ is, up to homeomorphism, the space that we called $L(p;q, 1)$ in the last section.

Consider the 3-sphere S^3 as the one-point compactification of R^3. Let Σ_1 be the unit circle in $R^2 \times 0$ and let $\Sigma_2 = (z\text{-axis}) \cup \{\infty\}$. Then S^3 can be seen as $\Sigma_1 * \Sigma_2$ by viewing it as the union of a suitable family of curved "cones" $v*\Sigma_1$ as v varies over Σ_2.

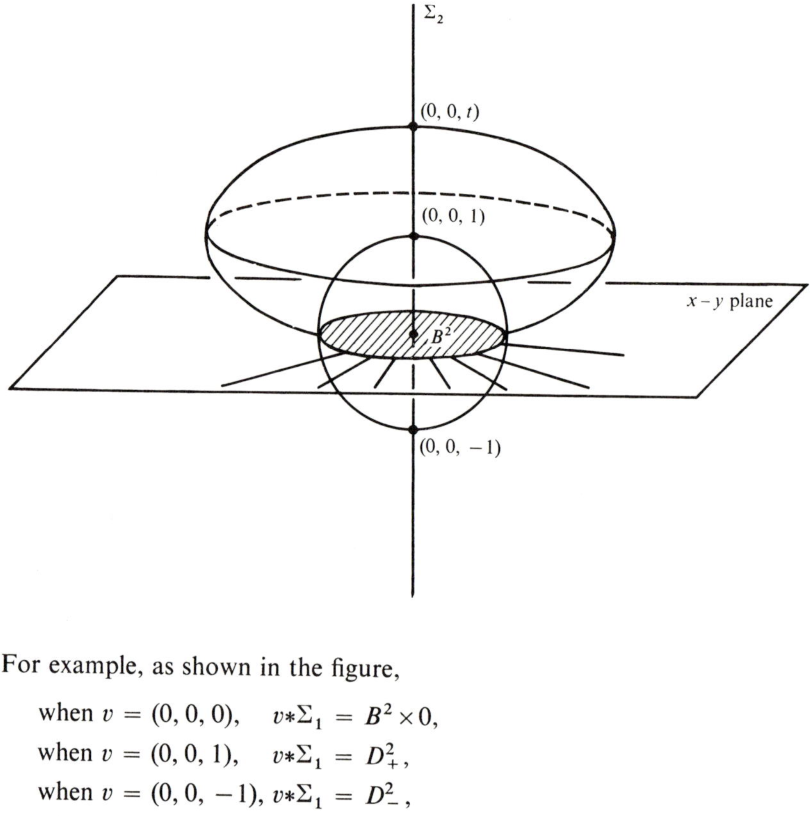

For example, as shown in the figure,

$$\begin{aligned}
&\text{when } v = (0, 0, 0), && v*\Sigma_1 = B^2 \times 0,\\
&\text{when } v = (0, 0, 1), && v*\Sigma_1 = D^2_+,\\
&\text{when } v = (0, 0, -1), && v*\Sigma_1 = D^2_-,\\
&\text{when } v = \infty, && v*\Sigma_1 = \{(x, y, 0) | x^2+y^2 \geq 1\},
\end{aligned}$$

and when $v = (0, 0, t)$, $|t| > 1$, $v*\Sigma_1$ looks like a turban. Each of these "cones" is gotten by rotating an arc from v to $(0, 1, 0)$, which lies in the y–z plane, about the z-axis, the rotated arcs giving us the cone lines.

Now let $R_1 : \Sigma_1 \to \Sigma_1$ be rotation through $2\pi q/p$ radians. Break Σ_2 into p line segments, one of which is the finite line segment from $(0, 0, -1)$ to $(0, 0, 1)$ and one of which is an infinite line segment which has ∞ as an interior point. Let $R_2 : \Sigma_2 \to \Sigma_2$ be the simplicial isomorphism which shifts each vertex to the next higher one, except that the highest vertex now becomes the lowest. Since every point of $S^3 - (\Sigma_1 \cup \Sigma_2)$ lies on a unique arc from Σ_1 to Σ_2 we may define $g = R_1 * R_2 : S^3 \to S^3$ by $g[z_1, z_2, t] = [R_1(z_1), R_2(z_2), t]$ where $[a, b, t]$ denotes the point which is $t \cdot L_{ab}$ units of arc-length along the arc from a to b, L_{ab} being the length of this arc.

If in this setting we let $G = \{1, g, g^2, \ldots, g^{p-1}\}$ it is clear that $S^3/G \approx L(p;q, 1)$ as defined in the last section. On the other hand, if $\pi: S^3 \to S^3/G$ is the quotient map it follows from the facts that $\pi g^k = \pi$ and $S^3 = \bigcup_k g^k(B^3)$, that $\pi(B^3) = S^3/G$. So S^3/G is homeomorphic to the quotient space of B^3 under the identifications induced by $\pi|B^3$. But this quotient space is precisely B^3/h since $g|D^2_- = h|D^2_-$ and $g(B^3) \cap B^3 = D^2_+$ and $g^k(B^3) \cap B^3 = \varnothing$ if $k \not\equiv \pm 1 (\text{mod } p)$. Hence

$$L(p;q, 1) \approx S^3/G \approx B^3/h = L_{p,q}.$$

§28. Cell structures and homology groups

When $p > 2$ we denote the vertices of Σ^1 by $v_j = e^{2\pi ij/p}$ and the 1-simplices by $I_j = [v_j, v_{j+1}]$, $0 \le j \le p-1$. When $p = 2$ the vertices are v_0, A, v_1, B (as in §26) and we set $I_0 = [v_0, A] \cup [A, v_1]$ and $I_1 = [v_1, B] \cup [B, v_0]$. For $0 \le i \le n-1$, the following simplicial subcomplexes of Σ^{2n-1} play an important role:

$$\Sigma^{2i-1} = \Sigma_1 * \Sigma_2 * \ldots * \Sigma_i \subset R^{2i}$$

$$E_j^{2i} = \Sigma^{2i-1} * v_j \quad (v_j \in \Sigma_{i+1})$$

$$E_j^{2i+1} = \Sigma^{2i-1} * I_j < \Sigma^{2i+1}.$$

(Here $\Sigma^{-1} \equiv \varnothing$, $E_j^0 \equiv v_j$, and $E_j^1 \equiv I_j$.) For example, B^3, D_-^2, D_+^2 of §27 correspond to E_j^3, E_j^2, E_{j+1}^2 where $v_j = (0, 0, -1)$ and $v_{j+1} = (0, 0, 1)$.

The E_j^k are closed cells—in fact *p.l.* balls—which give a CW decomposition of Σ^{2n-1} with cells $e_j^k = \mathring{E}_j^k$ $(0 \le j \le p-1, 0 \le k \le 2n-1)$. Elementary facts about joins imply that

(1) $\partial E_j^{2i} = \Sigma^{2i-1}$

(2) $\partial E_j^{2i+1} = E_j^{2i} \cup E_{j+1}^{2i}$

(3) $E_j^{2i} \cap E_k^{2i} = \Sigma^{2i-1}$, if $j \neq k$

$$E_j^{2i+1} \cap E_k^{2i+1} = \begin{cases} \Sigma^{2i-1} & \text{if } j-k \neq 0, \pm 1 \ (\text{mod } p) \\ E_{j+1}^{2i} & \text{if } k = j+1 \ (\text{mod } p), p \neq 2 \\ E_0^{2i} \cup E_1^{2i} & \text{if } j = 0, k = 1, p = 2. \end{cases}$$

Let us orient the balls E_j^k—think of them for the moment as simplicial chains—by stipulating that $E_j^0 = v_j$ is positively oriented and, inductively, that then E_j^{2i+1} is oriented $(i \ge 0)$ so that $\partial E_j^{2i+1} = E_{j+1}^{2i} - E_j^{2i}$, Σ^{2i+1} is oriented so that $E_j^{2i+1} \subsetneq \Sigma^{2i+1}$ is orientation preserving, and E_i^{2i+2} is oriented so that $\partial E_j^{2i+2} = \Sigma^{2i+1} = E_0^{2i+1} + E_1^{2i+1} + \ldots + E_{p-1}^{2i+1}$.
The orientations of the E_j^k naturally determine basis elements for the cellular chain complex $C_k(\Sigma^{2n-1})$ determined by this CW structure and we shall also use e_j^k to denote these basis elements (rather than $\langle \varphi_j^k \rangle$ for some characteristic map φ_j^k, as we did earlier).

Now view Σ^{2n-1} as the universal covering space of $L(p;q_1,\ldots,q_n) = \Sigma^{2n-1}/G$. It is natural, as we shall see shortly, to denote $e_0^k = \tilde{e}_k$ ($0 \le k \le 2n-1$). If $g = R_1 * R_2 * \ldots * R_n$, as before, notice that $g^t|\Sigma^{2i-1}: \Sigma^{2i-1} \to \Sigma^{2i-1}$ is an orientation preserving simplicial isomorphism (since it's homotopic to the identity) such that $g^t|E_0^{2i} = (g^t|\Sigma^{2i-1}) * (g^t|v_0)$ and $g^t|E_0^{2i+1} = (g^t|\Sigma^{2i-1}) * (g^t|I_0)$. Thus g^t takes oriented cells isomorphically in an orientation preserving manner to oriented cells and the basic cellular chains satisfy

$$e_j^k = g^t\tilde{e}_k \text{ where } tq_j \equiv j \,(\text{mod } p) \;[t \text{ exists because } (q_j, p) = 1]$$

$$\partial\tilde{e}_{2i+1} = g^{r_{i+1}}\tilde{e}_{2i} - \tilde{e}_{2i}, \text{ where } r_{i+1}q_{i+1} \equiv 1 \,(\text{mod } p)$$

$$(*) \qquad \partial\tilde{e}_{2i} = \tilde{e}_{2i-1} + g\tilde{e}_{2i-1} + \ldots + g^{p-1}\tilde{e}_{2i-1}$$

$$\partial g = g\partial : C^k(\Sigma^{2n-1}) \to C^k(\Sigma^{2n-1})$$

$L(p;q_1,\ldots,q_n)$ obtains a natural CW structure, with exactly one cell in each dimension from the cell structure on Σ^{2n-1} via the projection $\pi: \Sigma^{2n-1} \to L(p;q_1,\ldots,q_n)$. The cells are the sets $e_k = \pi(\tilde{e}_k) = \pi(e_j^k)$, $(0 \le j < p,\ 0 \le k \le 2n-1)$ with characteristic maps $\pi|\,E_0^k: E_0^k \to L(p;q_1,\ldots,q_n)$. The orientation we have chosen for $\tilde{e}_k$ induces an orientation for e_k. Or, more technically speaking, the chain map induced by the cellular map π takes the basis element $\tilde{e}_k$ of $C_k(\Sigma^{2n-1})$ to a basis element, denoted e_k, of $C_k(L(p;q_1,\ldots,q_n))$. To compute $H_*(L(p;q_1,\ldots,q_n))$ we simply note that

$$\begin{aligned} \partial e_{2i} &= \partial\pi(\tilde{e}_{2i}) = \pi\partial(\tilde{e}_{2i}) \\ &= \pi(\tilde{e}_{2i-1} + g\tilde{e}_{2i-1} + \ldots + g^{p-1}\tilde{e}_{2i-1}) = pe_{2i-1} \\ \partial e_{2i+1} &= \pi\partial(\tilde{e}_{2i+1}) = \pi(g^{r_{i+1}}\tilde{e}_{2i} - \tilde{e}_{2i}) = 0. \end{aligned}$$

Thus the cellular chain complex is

$$0 \to C_{2n-1} \xrightarrow{0} C_{2n-2} \xrightarrow{\times p} C_{2n-3} \xrightarrow{0} \cdots \xrightarrow{\times p} C_1 \xrightarrow{0} C_0 \to 0.$$

Hence the homology groups of $L(p;q_1,\ldots,q_n)$ with integral coefficients are

$$\begin{aligned} H_{2n-1} &= \mathbb{Z} \\ H_{2i-1} &= \mathbb{Z}_p, \quad 1 \le i < n \\ H_{2i} &= 0, \qquad i > 0 \\ H_0 &= \mathbb{Z}. \end{aligned}$$

Since the sphere is the universal covering space of the lens space and G is the group of covering homeomorphisms, $\pi_1 L(p;q_1,\ldots,q_n) = \mathbb{Z}_p$ and $\pi_i L(p;q_1,\ldots,q_n) = \pi_i S^{2n-1}$ for $i \neq 1$.

The preceding discussion shows that the different $(2n-1)$-dimensional lens spaces determined by a fixed p all have the same homology and homotopy groups. Nevertheless we shall show in the next section that they do not all have the same homotopy type.

§29. Homotopy classification[21]

We suppose throughout this section that $p \geq 2$ is given and that $(q_1, \ldots, q_n)$, $(q'_1, \ldots, q'_n)$ are given n-tuples of integers relatively prime to p. Let R_j and R'_j be rotations of Σ_j through q_j and q'_j notches respectively. Set $g = R_1 * \ldots * R_n$, $g' = R'_1 * \ldots * R'_n$ and let G, G' be the cyclic groups (of order p) generated by g, g'. Denote $L = L(p; q_1, \ldots, q_n) = \Sigma^{2n-1}/G$ and $L' = L(p; q'_1, \ldots, q'_n) = \Sigma^{2n-1}/G'$ with quotient maps π and π'. Also set $e_k = \pi(\tilde{e}_k)$ and $e'_k = \pi'(\tilde{e}_k)$, where $\tilde{e}_k$ is as in §28.

Definition: If $f: \Sigma^{2n-1} \to \Sigma^{2n-1}$ is a map and if $h \in G'$ then f is (g, h)-*equivariant* iff $fg = hf$. Two (g, h)-equivariant maps f_0 and f_1 are *equivariantly homotopic* if there is a homotopy $\{f_t\}: \Sigma^{2n-1} \to \Sigma^{2n-1}$ such that f_t is (g, h)-equivariant for all t.

Exercise: A map $f: \Sigma^{2n-1} \to \Sigma^{2n-1}$ is (g, h)-equivariant for at most one h.

The relationship between equivariant maps of Σ^{2n-1} and maps between lens spaces is given by the next two theorems.

(29.1) (a) *If $h \in G'$ and F is a (g, h)-equivariant map of Σ^{2n-1}, then F covers a well-defined map $f: L \to L'$.*

(b) *If $\{F_t\}$ is a homotopy of Σ^{2n-1}, where each F_t is (g, h_t)-equivariant for some $h_t \in G'$, then $h_t = h_0$ for all t and $\{F_t\}$ covers a well-defined homotopy $\{f_t\}: L \to L'$.*

PROOF: (a) Define f so that the following diagram commutes

$$\begin{array}{ccc} \Sigma^{2n-1} & \xrightarrow{\ F\ } & \Sigma^{2n-1} \\ \downarrow{\scriptstyle \pi} & & \downarrow{\scriptstyle \pi'} \\ L & \overset{f}{\dashrightarrow} & L' \end{array}$$

It is well-defined because $\pi(x) = \pi(y)$ implies $g^k(x) = y$ for some k and, hence, that $\pi' F(y) = \pi' F g^k(x) = \pi' h^k F(x) = \pi' F(x)$.

(b) Write $h_0 = h$ and let $S = \{t \in [0, 1] | h_t \neq h\}$. Suppose, contrary to our claim, that $S \neq \varnothing$. Set $u = \text{g.l.b.}\,(S)$. We first note that $u \notin S$. To see this, choose a sequence $t_i \to u$ such that $t_i \leq u$ and $h_{t_i} = h$. Fix a point z and note that $h_u^{-1}(F_u g(z)) = h_u^{-1}(h_u F_u(z)) = F_u(z) = \lim_i h^{-1} h_{t_i} F_{t_i}(z) = \lim_i h^{-1} F_{t_i} g(z) = h^{-1}(F_u g(z))$. Thus h_u^{-1} and h^{-1} agree at a point. So $h_u = h$ and $u \notin S$.

Now let t_i be a sequence in S such that $t_i \to u$. So $h_{t_i} \neq h$. Fix $z \in \Sigma^{2n-1}$. If V is a neighborhood of $F_u g(z)$ such that $\pi' | V$ is one-one, then $h^{-1}(V) \cap h_{t_i}^{-1}(V) = \varnothing$ for all i. But $h^{-1} F_u g(z) = F_u(z) = \lim_i h_{t_i}^{-1} F_{t_i} g(z)$ and we have a contradiction, since $F_{t_i} g(z)$ eventually lies in V, so $h_{t_i}^{-1} F_{t_i} g(z)$

---

21 The material in this section is essentially that of DERHAM's exposition [KERVAIRE-MAUMARY-DERHAM; p. 96–101].

eventually lies outside the neighborhood $h^{-1}(V)$ of $h^{-1}F_u g(z)$. Therefore S is empty and we see that each F_t is (g, h)-equivariant.

From (a) each F_t covers a map $f_t: L \to L'$, and the resultant function $f: L \times I \to L'$ is continuous since

$$\begin{array}{ccc} \Sigma^{2n-1} \times I & \xrightarrow{F} & \Sigma^{2n-1} \\ \downarrow{\scriptstyle \pi \times 1} & & \downarrow{\scriptstyle \pi'} \\ L \times I & \xrightarrow{f} & L' \end{array}$$

is a commutative diagram and $\pi \times 1$ is a closed map. □

(29.2) *If $f: L \to L'$, and if $f_\#: G \to G'$ comes from the induced map of fundamental groups (as in* (3.16)) *then the map $f_\#$ is independent of all possible choices of base points. In fact, if $h \in G'$ the following are equivalent assertions.*

(1) $f_\#(g) = h$.

(2) *Any map $\tilde{f}: \Sigma^{2n-1} \to \Sigma^{2n-1}$ which covers f is (g, h)-equivariant.*

(3) *There is some map $\tilde{f}$ covering f which is (g, h)-equivariant.*

Moreover, if $f_0 \simeq f_1: L \to L'$ (free homotopy) then $f_{0\#} = f_{1\#}$.

REMARK: Because of this theorem almost no references to base points will be made in this chapter, and statements about what the map $f: L \to L'$ does on fundamental groups will be given in terms of the map $f_\#: G \to G'$.

PROOF: If $\tilde{f}$ covers f, choose points x, y and points $\tilde{x}$, $\tilde{y}$ covering them such that $\tilde{f}(\tilde{x}) = \tilde{y}$, $f(x) = y$. Then, by (3.16), $f_\#(g) \circ \tilde{f} = \tilde{f}g$. Thus every lift $\tilde{f}$ is (g, h)-equivariant for some h—namely $h = f_\#(g)$, where $f_\# = \theta(y, \tilde{y}) \circ$ (induced map on π_1) $\circ\, \theta(x, \tilde{x})^{-1}$.

If $\tilde{f}$ is (g, h)-equivariant and $\hat{f}$ is another lift, then $\hat{f} = k\tilde{f}$ for some $k \in G'$ and we have $\hat{f}g = k\tilde{f}g = kh\tilde{f} = hk\tilde{f} = h\hat{f}$. Thus (3) ⇒ (2). Trivially then (3) ⇔ (2).

If $\tilde{f}$ and $\hat{f}$ are two lifts, each giving rise to an $f_\#$, one with $f_\#(g) = h$ and the other with $f_\#(g) = h'$, then as noted above $\tilde{f}$ and $\hat{f}$ are (g, h)- and (g, h')-equivariant. Since (3) ⇒ (2), we see that $h = h'$, so in fact $f_\#$ is well-defined, independently of all choices.

From the above, the equivalence (1) ⇔ (2) ⇔ (3) is now obvious.

Finally, if $\{f_t\}: L \to L'$ is a homotopy from f_0 to f_1, let $\{\tilde{f}_t\}$ be a lift to a homotopy of Σ^{2n-1}. Each $\tilde{f}_t$ is (g, h_t)-equivariant, where $f_{t\#}(g) = h_t$. But by (29.1b), $h_t = h_0$ for all t. Hence $f_{0\#} = f_{1\#}$. □

In the light of (29.1) and (29.2) we shall first derive some results about equivariant maps of Σ^{2n-1} and then interpret these as results concerning homotopy classes of maps $L \to L'$.

(29.3) *If f_0 and f_1 are any two (g, h)-equivariant maps of Σ^{2n-1} then degree $f_0 \equiv$ degree f_1 (mod p). If, in fact, degree $f_0 =$ degree f_1 then f_0 and f_1 are equivariantly homotopic.*

PROOF: Let Σ^{2n-1} be viewed as a cell complex—call it K—as in §28, and let $P = K \times I$, a $2n$-dimensional complex. We claim first that there is a map $F: P^{2n-1} \to \Sigma^{2n-1}$ (where P^i denotes the i-skeleton of P) such that

$$F_t = f_t \quad (t = 0, 1)$$

$$F_t g(x) = hF_t(x) \quad \text{for all} \quad (x, t) \in P^{2n-1}$$

where $F_t = F|(K \times \{t\}) \cap P^{2n-1}$.

The function F is constructed inductively over the complexes $P_i = P^i \cup K \times \{0, 1\}$. We set $F^0|P_0 = f_0 \cup f_1$. Suppose that $F^i|P_i$ has been defined and satisfies the equivariant property, where $0 \le i < 2n-1$. In particular $F^i|\partial(E_0^i \times I)$ has been defined. Since $\dim \partial(E_0^i \times I) < 2n-1$, there is an extension $F^{i+1}: E_0^i \times I \to \Sigma^{2n-1}$. Now define $F^{i+1}: E_j^i \times I \to \Sigma^{2n-1}$ for $j \ge 1$ by

$$F_t^{i+1}(x) = h^{-q}(F_t^{i+1}|E_0^i)g^q(x) \quad \text{where } g^q(E_j^i) = E_0^i$$

This is well-defined on all of P_{i+1} because, since $(q_j, p) = 1$, there is exactly one $q \pmod p$ with $g^q(E_j^i) = E_0^i$, and because, if $(x, t) \in \partial(E_j^i \times I) \subset P_i$ the induction hypothesis $F_t^i(y) = h^{-1}F_t^i g(y)$ applied to $y = g^k(x)$, $k = 0, 1, \ldots, q-1$, implies

$$F_t^i(x) = h^{-1}F_t^i(g(x)) = h^{-2}F_t^i g^2(x) = \ldots = h^{-q}F_t^i g^q(x)$$
$$= h^{-q}(F_t^i|E_0^i)g^q(x) = F_t^{i+1}(x).$$

F^{i+1} has the equivariant property on P_{i+1} because if $x \in E_j^i$ then $g(x) \in E_k^i$ where $g^{q-1}(E_k^i) = E_0^i$. Hence

$$F_t^{i+1}g(x) = h^{-(q-1)}(F_t^{i+1}|E_0^i)g^{q-1}(g(x)) = hF_t^{i+1}(x)$$

Setting $F = F^{2n-1}$, the proof of the claim is completed.

As in §28 we assume that Σ^{2n-1} and the E_j^{2n-1} are oriented so that each inclusion $E_j^{2n-1} \subset \Sigma^{2n-1}$ $(0 \le j < p)$ is orientation preserving. Give $\Sigma^{2n-1} \times I$ the product orientation so that $\partial(\Sigma^{2n-1} \times I) = (\Sigma^{2n-1} \times 1) - (\Sigma^{2n-1} \times 0)$. This induces an orientation on each $E_j^{2n-1} \times I$ and hence also on $\partial(E_j^{2n-1} \times I)$. Let $F_j = F|\partial(E_j^{2n-1} \times I): \partial(E_j^{2n-1} \times I) \to \Sigma^{2n-1}$. Since the range and domain of F_j are oriented $(2n-1)$-spheres the degree of F_j is well-defined. Notice that $F_j|E_{j+1}^{2n-2} \times I = F_{j+1}|(E_{j+1}^{2n-2} \times I)$ and, thinking of cellular chains, that $\partial(E_j^{2n-1} \times I) = (E_j^{2n-1} \times 1) - (E_j^{2n-1} \times 0) + (E_{j+1}^{2n-2} \times I) - (E_j^{2n-2} \times I)$. Then it is an elementary exercise (or use [Hilton-Wylie, II.1.30] and the Hurewicz homomorphism) that

$$\sum_j \deg F_j = \deg f_1 - \deg f_0.$$

But $F_j = h^{-q}F_0(g^q \times 1)$ where $g^q \times 1$ takes $\partial(E_j^{2n-1} \times I)$ onto $\partial(E_0^{2n-1} \times I)$ by a degree 1 homeomorphism, and $h^{-q}: \Sigma^{2n-1} \to \Sigma^{2n-1}$ is also of degree 1. Hence $\deg F_j = \deg F_0$. So $\deg f_1 - \deg f_0 = \sum_j \deg F_j = p \cdot \deg F_0 \equiv 0 \pmod p$, proving the first assertion of our theorem.

If, in fact, $\deg f_1 = \deg f_0$ the last sentence shows that $\deg F_0 = 0$. By Brouwer's theorem F_0 may then be extended over $E_0^{2n-1} \times I$ and so, as above, F may be extended to $P_{2n} = \Sigma^{2n-1} \times I$ by stipulating that $F_t|E_j^{2n-1} = h^{-q}(F_t|E_0)g^q$. This extension is the desired equivariant homotopy. □

The question of which residue class mod p it is which is determined by the degree of an equivariant map, and whether all numbers in the residue class can be realized, is answered by

(29.4) *If* $(d, a) \in \mathbb{Z} \times \mathbb{Z}$ *then there is a* (g, g'^a) *equivariant map* $f: \Sigma^{2n-1} \to \Sigma^{2n-1}$ *of degree* d *if and only if* $d \equiv a^n r_1 \ldots r_n q_1' \ldots q_n' \pmod p$ *where* $r_j q_j \equiv 1 \pmod p$.

PROOF: For each j fix an r_j with $r_j q_j \equiv 1 \pmod p$. An equivariant map of degree $d_0 = a^n r_1 \ldots r_n q_1' \ldots q_n'$ can be constructed by simply wrapping each Σ_j about itself $r_j a q_j'$ times and taking the join of these maps. To make this precise we use the notation of complex numbers. Define $T: \Sigma^1 \to S^1$ by $T(z) = (z/|z|)$. Think of the rotations R_j, R_j' as acting on all R^2. These commute with T. Let $m_j(z) = z^{r_j a q_j'}$ (z complex) and define $f_j: \Sigma^1 \to \Sigma^1$ by $f_j = T^{-1} m_j T$. Then we claim that $f = f_1 * f_2 * \ldots * f_n: \Sigma^{2n-1} \to \Sigma^{2n-1}$ is a (g, g'^a) equivariant map of degree d_0.

The equivariance will follow immediately once we know that $(R_j')^a f_j = f_j R_j$, and this is true because

$$\begin{aligned}(R_j')^a f_j(z) &= (R_j')^a T^{-1} m_j T(z) = T^{-1}(R_j')^a m_j T(z) \\ &= T^{-1}(R_j')^{a r_j q_j} m_j T(z) = T^{-1}(e^{2\pi i \cdot q_j' a r_j q_j/p} \cdot T(z)^{a r_j q_j'}) \\ &= T^{-1}((R_j T(z))^{a r_j q_j'}) = T^{-1} m_j R_j T(z) \\ &= T^{-1} m_j T R_j(z) = f_j R_j(z)\end{aligned}$$

That the degree of f is indeed d_0 can be seen directly by counting. {We give the argument when $p \neq 2$. Slight adjustments in notation are necessary when $p = 2$.} There is a subdivision $\hat{\Sigma}_j$ of Σ_j into $|p r_j a q_j'|$ 1-simplexes—say $[v_k, v_{k+1}]$ gets divided into $v_k = v_{k,0}, v_{k,1}, \ldots, v_{k,|r_j a q_j'|} = v_{k+1}$—such that f_j takes simplexes of $\hat{\Sigma}_j$ homeomorphically onto simplexes of Σ_j; i.e., $f_j[v_{k,b}, v_{k,b+1}] = [v_s, v_{s\pm1}]$ for some s, the sign agreeing with the sign of $r_j a q_j'$. Let $\hat{\Sigma}^{2n-1} = \hat{\Sigma}_1 * \ldots * \hat{\Sigma}_n$. Then one generator α of the simplicial cycles $Z_{2n-1}(\hat{\Sigma}^{2n-1})$ is the chain which is the sum of $(2n-1)$-simplexes of the form

$$[v_{k,a}, v_{k,a+1}] * [v_{\ell,b}, v_{\ell,b+1}] * \ldots * [v_{m,c}, v_{m,c+1}] < \hat{\Sigma}_1 * \hat{\Sigma}_2 * \ldots \hat{\Sigma}_n$$

Such a simplex goes under f to $\pm$ a typical simplex in the similarly chosen generator β of $Z_{2n-1}(\Sigma^{2n-1})$, the sign being the product of the signs of $r_j a q_j'$; i.e. the sign of d_0. Counting the possible simplexes involved, one concludes that $f_*(\alpha) = d_0 \beta$ as claimed.

From (29.3) and the fact that there is an equivariant map of degree d_0 we conclude that if d is the degree of any (g, g'^a)-equivariant map then $d \equiv d_0 \,(= a^n r_1 \ldots r_n q_1' \ldots q_n') \pmod p$. Conversely suppose that $d = d_0 + Np$ where d_0 is the degree of a (g, g'^a)- equivariant map f. We modify f as follows to get a (g, g'^a)-equivariant map of degree d:

Let Q_0 be a round closed ball in the interior of a top dimensional simplex of E_0^{2n-1}. Using a coordinate system with origin at the center of Q_0 we write

$$Q_0 = \{tx \mid x \in \mathrm{Bdy}\, Q_0, 0 \le t \le 1\};\ \tfrac{1}{2}\, Q_0 = \{tx \mid 0 \le t \le \tfrac{1}{2}\}.$$

$$Q_j = g^j(Q_0),\ (0 \le j \le p); \qquad \tfrac{1}{2}\, Q_j = g^j(\tfrac{1}{2}\, Q_0).$$

Define $h\colon \Sigma^{2n-1} \to \Sigma^{2n-1}$ by

(A) $h|(\Sigma^{2n-1} - \bigcup_j \mathrm{Int}\ Q_j) = f|(\Sigma^{2n-1} - \bigcup_j \mathrm{Int}\ Q_j)$

(B) $h(tx) = f((2t-1)x)$, if $\tfrac{1}{2} \le t \le 1$

$h\colon (\tfrac{1}{2}\, Q_0, \mathrm{Bdy}(\tfrac{1}{2}\, Q_0)) \to (\Sigma^{2n-1}, f(0))$ is any map of degree N you like

(C) $h|Q_j = (g'^a)^j\,(h|Q_0)g^{-j}$.

The check that h is equivariant is straightforward, as in the proof of (29.3). To check the degree of h, let $C = \Sigma^{2n-1} - \bigcup_j \mathrm{Int}\,(\tfrac{1}{2}Q_j)$. Consider $(h|C)_*\colon H_n(C, \mathrm{Bdy}\ C) \to H_n(\Sigma^{2n-1}, f(0))$ and $(h|\tfrac{1}{2}Q_j)_*\colon H_n(\tfrac{1}{2}\ Q_j, \mathrm{Bdy}\ \tfrac{1}{2}\ Q_j) \to H_n(\Sigma^{2n-1}, f(0))$. It is an elementary exercise (or use [Hilton-Wylie, II.1.31] and the Hurewicz homomorphism) that $\deg h = \deg(h|C) + \sum_j \deg(h|\tfrac{1}{2}\ Q_j)$. But g^{-j} takes Q_j in an orientation preserving manner onto Q_0 and $(g'^a)^j$ is just a rotation of Σ^{2n-1}. Hence $(h|\tfrac{1}{2}Q_j)_*(\text{generator}) = (h|\tfrac{1}{2}Q_0)_*(\text{generator}) = N\cdot(\text{generator})$. Since clearly, $\deg f = \deg\,(h|C)$, we have $\deg h = \deg f + \sum_j N = d_0 + Np$, as desired. □

We now turn to the interpretation of these equivariant results as results about maps between lens spaces.

(29.5) *Suppose that* $L = L(p; q_1, \ldots, q_n)$ *and* $L' = L(p; q_1', \ldots, q_n')$ *are oriented by choosing* e_{2n-1} *and* e'_{2n-1} *as generators of* H_{2n-1} (see §28).

(A) *If* $f_0, f_1\colon L \to L'$ *then* $[f_0 \simeq f_1] \Leftrightarrow [\deg f_0 = \deg f_1$ *and* $f_{0\#} = f_{1\#}\colon G \to G']$. (See the Remark following (29.2).)

(B) *If* $(d, a) \in \mathbb{Z} \times \mathbb{Z}$ *then* [*there is a map* $f\colon L \to L'$ *such* that $\deg f = d$ and $f_\#(g) = g'^a] \Leftrightarrow [d \equiv a^n r_1, \ldots, r_n q_1', \ldots, q_n' \pmod p$, where $r_j q_j \equiv 1 \pmod p$].

(C) If $f\colon L \to L'$ then f is a homotopy equivalence $\Leftrightarrow \deg f = \pm 1$.

PROOF: (A) Suppose that $\deg f_0 = \deg f_1$ and $f_{0\#}(g) = f_{1\#}(g) = h$. Choose lifts $\tilde{f}_i\colon \Sigma^{2n-1} \to \Sigma^{2n-1}$, $i = 0, 1$. These are both (g, h)-equivariant by (29.2). They have the same degree because if $z = \sum_{i=0}^{p-1} g^i \tilde{e}_{2n-1} = \sum_{i=0}^{p-1} g'^i \tilde{e}_{2n-1}$ is chosen as basic cycle for $C_{2n-1}(\Sigma^{2n-1})$ then $\pi_*(z) = \pi'_*(z) = p\cdot e_{2n-1}$, and the commutative diagram

$$\begin{array}{ccc}
\Sigma^{2n-1} & \xrightarrow{\ \tilde{f}_i\ } & \Sigma^{2n-1} \\
{\scriptstyle (\deg p)}\downarrow{\scriptstyle \pi} & & {\scriptstyle (\deg p)}\downarrow{\scriptstyle \pi'} \\
L & \xrightarrow[\ f_i\ ]{} & L'
\end{array}$$

shows that $\deg \tilde{f}_0 = \deg f_0 = \deg f_1 = \deg \tilde{f}_1$. Hence, by (29.3) $\tilde{f}_0$ and $\tilde{f}_1$

are equivariantly homotopic. So by (29.1), $f_0 \simeq f_1$. The converse is trivial using (29.2).

(B) Given f such that $\deg f = d$ and $f_\#(g) = g'^a$, let $\tilde{f}\colon \Sigma^{2n-1} \to \Sigma^{2n-1}$ lift f. Then $\tilde{f}$ is a (g, g'^a)-equivariant map which, by the argument for (A), has degree d. Hence by (29.4), $d \equiv a^n r_1 \ldots r_n q'_1 \ldots q'_n \pmod{p}$.

Conversely, if d satisfies this congruence, there is, by (29.4), a (g, g'^a) equivariant map $F\colon \Sigma^{2n-1} \to \Sigma^{2n-1}$ of degree d. By (29.1) and (29.2), F covers a map $f\colon L \to L'$ such that $f_\# (g) = (g')^a$. As before degree f = degree $F = d$.

(C) Suppose that $f\colon L \to L'$ has degree ± 1. Assume $f_\#(g) = g'^a$. Then

$$\deg f = \pm 1 \equiv a^n r_1 \ldots r_n q'_1 \ldots q'_n \pmod{p}$$

i.e.,

$$(*)\quad a^n \equiv \pm q_1 \ldots q_n r'_1 \ldots r'_n \pmod{p}, \quad \text{where } r'_j q'_j \equiv 1 \pmod{p}.$$

Thus $(a, p) = 1$ and we may choose b such that $ab \equiv 1 \pmod{p}$. From $(*)$

$$1 \equiv b^n a^n \equiv \pm b^n q_1 \ldots q_n r'_1 \ldots r'_n \pmod{p}$$

Hence, by (B), there is a map $\bar{f}\colon L' \to L$ with $\deg \bar{f} = \deg f = \pm 1$ and $\bar{f}_\#(g') = (g')^b$. Then $\deg(\bar{f} f) = 1$ and $(\bar{f} f)_\# (g) = g^{ab} = g$. So by (A), $\bar{f} f \simeq 1_L$. Similarly $f \bar{f} \simeq 1_{L'}$ and f is a homotopy equivalence. □

Notice that, when $p = 2$, $L = L' = L(2; 1, 1, \ldots, 1) = \mathbb{R}P^{2n-1}$ (real projective space). The preceding proposition tells us that there are exactly two self-homotopy equivalences of $\mathbb{R}P^{2n-1}$, one of degree $+1$ and one of degree -1. All other cases are given by the following **classification of homotopy equivalences**.

(29.6) Suppose that $L = L(p; q_1, \ldots, q_n)$ and $L' = L(p; q'_1, \ldots, q'_n)$ *where* $p > 2$, *and that* $r'_j q'_j \equiv 1 \pmod{p}$ *for all* j. *Let* $\mathscr{E}[L, L']$ *denote the set of equivalence classes under homotopy of homotopy equivalences* $f\colon L \to L'$. *Then there is a bijection*

$$\varphi\colon \mathscr{E}[L, L'] \to \{a \mid 0 < a < p \text{ and } a^n \equiv \pm q_1 \ldots q_n r'_1 \ldots, r'_n \pmod{p}\}$$

given by $\varphi[f] = a$ *if* $f_\# (g) = g'^a$. *Moreover, if* $\varphi[f] = a$ *then degree* $f = \pm 1$ *where the sign agrees with that above.*

PROOF: A straightforward application of (29.5). □

The following applications of (29.6) are left as exercises in arithmetic.

(I) *$\mathscr{E}[L, L]$ is isomorphic to the group consisting of those units a in the ring $\mathbb{Z}_p$ such that $a^n \equiv \pm 1 \pmod{p}$, provided $p \neq 2$.*

(II) *Any homotopy equivalence of $L_{7,q}$ onto itself is of degree $+1$.* Thus $L_{7,q}$ admits no orientation reversing self-homeomorphism. Such a manifold is called *asymmetric*.

(III) *$L_{p,q}$ and $L_{p,q'}$ have the same homotopy type $\Leftrightarrow$ there is an integer b such that $qq' \equiv \pm b^2 \pmod{p}$.* Thus we have the examples:

$$L_{5,1} \not\simeq L_{5,2}$$

$$L_{7,1} \simeq L_{7,2}$$

where "$\simeq$" denotes homotopy equivalence. We shall show in the next section that $L_{7,1}$ and $L_{7,2}$ do not have the same simple-homotopy type. (Compare (24.4).)

§30. Simple-homotopy equivalence of lens spaces

The purpose of this section is to prove

(30.1) *Let $L = L(p;q_1, \ldots, q_n)$ and $L' = L(p;q_1', \ldots, q_n')$ and suppose that f: $L \to L'$ is a simple-homotopy equivalence. If $f_\#(g) = (g')^a$ (as explained by* (29.2)) *then there are numbers $\varepsilon_i \in \{+1, -1\}$ such that $(q_1, \ldots, q_n)$ is equal (mod p) to some permutation of $(\varepsilon_1 a q_1', \varepsilon_2 a q_2', \ldots, \varepsilon_n a q_n')$.*

Our proof will not be totally self-contained in that we shall assume the following number-theoretic result. (For a proof see [KERVAIRE-MAUMARY-DERHAM, p. 1–12].)

Franz' theorem: Let $S = \{j \in \mathbb{Z} | 0 < j < p, \ (j, p) = 1\}$. Suppose that $\{a_j\}_{j \in S}$ is a sequence of integers, indexed by S, satisfying

(1) $\sum_{j \in S} a_j = 0$

(2) $a_j = a_{p-j}$

(3) $\prod_{j \in S} (\xi^j - 1)^{a_j} = 1$ for every p^{th} root of unity $\xi \neq 1$.

Then $a_j = 0$ for all $j \in S$.

PROOF OF (30.1): We give $\Sigma^{2n-1} = \tilde{L} = \tilde{L}'$ the cell structure of §28. Then $C(\tilde{L}')$ and $C(\tilde{L})_{f_\#}$ are $\mathbb{Z}(G')$-complexes with basis $\{\tilde{e}_k\}$ in dimension k and boundary operators gotten from equations (∗) on page 90. Denoting $\Sigma(x) = 1 + x + \ldots + x^{p-1}$, $C_j(\tilde{L}') = C_j'$, and $[C_j(\tilde{L})]_{f_\#} = C_j$, these complexes look like

$$C(\tilde{L}'): 0 \to C_{2n-1}' \xrightarrow{(g')^{r_n'} - 1} C_{2n-2}' \xrightarrow{\Sigma(g')} C_{2n-3}' \xrightarrow{(g')^{r_{n-1}'} - 1}$$
$$\cdots \xrightarrow{\Sigma(g')} C_1' \xrightarrow{(g')^{r_1'} - 1} C_0' \to 0$$

$$C(\tilde{L})_{f_\#}: 0 \to C_{2n-1} \xrightarrow{(g')^{ar_n} - 1} C_{2n-2} \xrightarrow{\Sigma(g'^a)} C_{2n-3} \to \cdots$$
$$\cdots \xrightarrow{\Sigma(g'^a)} C_1 \xrightarrow{(g')^{ar_1} - 1} C_0 \to 0.$$

Now invoke (22.8). Thus $0 = \tau(f) = \tau(\mathscr{C})$ where $\mathscr{C}$ is an acyclic $Wh(G')$-complex which fits into a based short exact sequence of $Wh(G')$-complexes

$$0 \to C(\tilde{L}') \to \mathscr{C} \to \bar{C}(\tilde{L}) \to 0$$

where $\bar{C}(\tilde{L})$ is the complex $C(\tilde{L})_{f_\#}$ shifted in dimension by one and with boundary operator multiplied by (-1).

It would be quite useful if the complexes in this last sequence were all acyclic. To achieve this we change rings. Suppose that ξ is any p^{th} root of unity other than 1. Let $\mathbb{C}$ be the complex numbers and let $h:\mathbb{Z}(G') \to \mathbb{C}$ by $h(\sum_j n_j(g')^j) = \sum_j n_j\xi^j$. Then, by the discussion at the end of §18—namely point 8. on page 61—we have a based short exact sequence

$$0 \to C(\tilde{L}')_h \to \mathscr{C}_h \to \bar{C}(\tilde{L})_h \to 0.$$

But now $C(\tilde{L}')_h$ and $\bar{C}(\tilde{L})_h$ are acyclic $(\mathbb{C}, \bar{G})$-complexes, by (18.1), where $\bar{G} = \{\pm\xi^j | j \in \mathbb{Z}\}$. [To apply (18.1) one must note that the fact that $(a, p) = 1$ implies that $\Sigma(g') = \Sigma(g'^a)$ and $(ar_j, p) = 1$ for all j.] Moreover, also by (18.1)

$$\tau(C(\tilde{L}')_h) = \tau\left\langle \prod_{k=1}^{n} (\xi^{r'_k} - 1\right\rangle \in K_{\bar{G}}(\mathbb{C})$$

$$\tau(\bar{C}(\tilde{L})_h) = -\tau\left\langle \prod_{k=1}^{n} (\xi^{ar_k} - 1)\right\rangle \in K_{\bar{G}}(\mathbb{C}).$$

The minus sign in the last equation comes from the shift in dimension. The change of sign of the boundary operator has no effect since $\tau(d+\delta) = \tau(-d-\delta)$. In this setting (17.2) and (18.2) yield

$$\begin{aligned} 0 = h_*(\tau\mathscr{C}) &= \tau(\mathscr{C}_h) \\ &= \tau(C(\tilde{L}')_h) + \tau(\bar{C}(\tilde{L})_h) \\ &= \tau\left\langle \prod_{k=1}^{n} (\xi^{r'_k} - 1)\right\rangle - \tau\left\langle \prod_{k=1}^{n} (\xi^{ar_k} - 1)\right\rangle. \end{aligned}$$

The determinants of these 1×1 matrices can only differ by a factor lying in $\bar{G}$ (using (10.6)). So we conclude

$$(*) \qquad \prod_{k=1}^{n} (\xi^{r'_k} - 1) = \pm\, \xi^s \prod_{k=1}^{n} (\xi^{ar_k} - 1),$$

for every p^{th} root of unity $\xi \neq 1$.

From here on its just a case of doing some manipulating to show that our theorem follows from equation (*) and the FRANZ Theorem. But without FRANZ' Theorem the reader should pause and do the

Exercise: *If $f: L_{7,2} \to L_{7,1}$ is a homotopy equivalence, so that, according to* (29.6), $f_\#(g) = g'^a$ *where* $a^2 \equiv 2$ *or* $a^2 \equiv 5$ (mod 7), *and if* $\xi = e^{2\pi i/7}$, *then*

$$|(\xi - 1)^2| \neq |(\xi^a - 1)(\xi^{4a} - 1)|.$$

Hence f is not a simple-homotopy equivalence.

Now simplify notation by writing $s_k = r'_k$, $t_k = ar_k$. Equation (*) gives, for every non-trivial p' *th* root of unity,

$$|\prod_{k=1}^{n} (\xi^{s_k}-1)|^2 = |\prod_{k=1}^{n} (\xi^{t_k}-1)|^2$$

or

$$(**) \qquad \prod_{k=1}^{n} (\xi^{s_k}-1)(\xi^{-s_k}-1) = \prod_{k=1}^{n} (\xi^{t_k}-1)(\xi^{-t_k}-1),$$

since $(\xi^{-d}-1)$ is the complex conjugate of $(\xi^{d}-1)$.

If $j \in S$, let S_j be the subsequence of $(s_1, -s_1, s_2, -s_2, \ldots, s_n, -s_n)$ consisting of those terms x such that $x \equiv j \pmod p$. Let m_j be the length of S_j. Similarly define the sequence T_j with length m'_j from the sequence $(t_1, -t_1, \ldots, t_n, -t_n)$. Since $(s_k, p) = 1$ implies $\pm s_k \equiv j \pmod p$ for some $j \in S$, and $i \neq j$ implies $S_i \cap Sj = \emptyset$, the sequence $(s_1, -s_1, \ldots, s_n, -s_n)$ is the disjoint union of the S_j. Hence $\sum_{j \in S} m_j = 2n$. Also the correspondence $x \mapsto -x$ gives a bijection from S_j to S_{p-j}, so $m_j = m_{p-j}$. Of course, similar equations hold for the m'_j. Let $a_j = m_j - m'_j$. Then

(1) $\sum_{j \in S} a_j = 2n - 2n = 0$

(2) $a_j = m_j - m'_j = m_{p-j} - m'_{p-j} = a_{p-j}$

(3) If $\xi \neq 1$ is any root of unity and if

$S_j = (\varepsilon_{j1} s_{j1}, \ldots, \varepsilon_{jm_j} s_{jm_j})$, $T_j = (\delta_{j1} t_{j1}, \ldots, \delta_{jm'_j} t_{jm'_j})$

with $\varepsilon_{j\alpha}, \delta_{j\beta} = \pm 1$, then

$$\begin{aligned}\prod_{j \in S} (\xi^j - 1)^{a_j} &= \prod_{j \in S} (\xi^j - 1)^{m_j} (\xi^j - 1)^{-m'_j} \\ &= \prod_{j \in S} \left[\left(\prod_{i=1}^{m_j} (\xi^{\varepsilon_{ji} s_{ji}} - 1)\right)\left(\prod_{i=1}^{m'_j} (\xi^{\delta_{ji} t_{ji}} - 1)\right)^{-1}\right] \\ &= \prod_{k=1}^{n} (\xi^{s_k}-1)(\xi^{-s_k}-1)(\xi^{t_k}-1)^{-1}(\xi^{-t_k}-1)^{-1} \\ &= 1 \text{ from } (**).\end{aligned}$$

Hence, by FRANZ' theorem, each $a_j = 0$ and $m_j = m'_j$. But, if $p \neq 2$, m_j is just the number of terms among $(r'_1, \ldots, r'_n)$ which are congruent to $\pm j$, mod p; and similarly for m'_j and $(ar_1, \ldots, ar_n)$. Hence under some reordering $r'_{i_1}, \ldots, r'_{i_n}$ we have

$$\varepsilon_{i_k} r'_{i_k} \equiv a r_k \pmod p, \ \varepsilon_{i_k} \in \{+1, -1\}, \ k = 1, 2, \ldots, n.$$

So $\quad \varepsilon_{i_k} q'_{i_k} \equiv a^{-1} q_k \pmod p$

and $\quad \varepsilon_{i_k} a q'_{i_k} \equiv q_k \pmod p$.

If $p = 2$, then $a = 1$ and $(q_1, \ldots, q_n) = (q'_1, \ldots, q'_n) = (1, 1, \ldots, 1) \pmod p$, so there was nothing to prove in the first place. □

§31. The complete classification

If $L = L(p;q_1, \ldots, q_n)$ and $L' = L(p;q'_1, \ldots, q'_n)$, the following assertions are equivalent:[22]

(A) *There is a number a and there are numbers $\varepsilon_j \in \{-1, 1\}$ such that $(q_1, \ldots, q_n)$ is a permutation of $(\varepsilon_1 a q'_1, \ldots, \varepsilon_n a q'_n)$ (mod p).*

(B) *L is simple-homotopy equivalent to L'.*

(C) *L is p.l. homeomorphic to L'.*

(D) *L is homeomorphic to L'.*

Moreover every simple-homotopy equivalence (and thus by [K-S] every homeomorphism) between lens spaces is homotopic to a p.l. homeomorphism.

PROOF: Clearly $(C) \Rightarrow (D)$. The implication $(D) \Rightarrow (B)$ is true because of (25.4) Thus the equivalence of (A)–(D) will follow from the equivalence of (A)–(C).

We have already proved [(25.3) and (30.1)] that $(C) \Rightarrow (B) \Rightarrow (A)$. To prove $(A) \Rightarrow (C)$, suppose that $q_i = \varepsilon_{j_i} a q'_{j_i}$ where $(j_1, \ldots, j_n)$ is a permutation of $(1, 2, \ldots, n)$. Think of Σ^{2n-1} as $\Sigma_1 * \Sigma_2 * \ldots * \Sigma_n$ where $\Sigma_i \subset \underbrace{0 \times \ldots \times R^2}_{i} \times \ldots \times 0$. Let $T_i : \Sigma_i \to \Sigma_{j_i}$ be the simplicial isomorphism given by

$$T_i\,(\underbrace{0, \ldots, 0, z}_{i}, \ldots, 0) = (\underbrace{0, 0, \ldots, z}_{j_i}, 0, \ldots, 0), \text{ if } \varepsilon_{j_i} = 1$$

$$T_i\,(\underbrace{0, \ldots, 0, z}_{i}, 0, \ldots, 0) = (\underbrace{0, 0, \ldots, 0, \bar{z}}_{j_i}, 0, \ldots, 0), \text{ if } \varepsilon_{j_i} = -1,$$

where $\bar{z}$ is the complex conjugate of z. With R and R' as in §26, $(R'_{j_i})^a(w) = (e^{2\pi i q'_{j_i} a/p} \cdot w)$ and $R_i\,(w) = (e^{2\pi i q_i/p} \cdot w)$; so it follows that $(R'_{j_i})^a T_i = T_i R_i$. Then the simplicial isomorphism $T(\sum_i \lambda_i z_i) = \sum_i \lambda_i T_i(z_i)$ is a (g, g'^a)-equivariant p.l. homeomorphism of Σ^{2n-1}. This induces a map $h: L \to L'$ via the diagram

$$\begin{array}{ccc} \Sigma^{2n-1} & \xrightarrow{\;T\;} & \Sigma^{2n-1} \\ \downarrow{\scriptstyle \pi} & & \downarrow{\scriptstyle \pi'} \\ L & \xrightarrow{\;h\;} & L' \end{array}$$

h is p.l. since π, π' and T are p.l., and h is a homeomorphism because T, being both equivariant and a homeomorphism, cannot take points in two different fibers into the same fiber. Thus (A) $\Rightarrow$ (C) as claimed.

---

22 Had we considered lens spaces as smooth manifolds, as at the end of §26, we could also add (E): *L is diffeomorphic to L'.*

Finally suppose that $f: L \to L'$ is a simple-homotopy equivalence with $f_{\#}(g) = g'^a$. By (30.1), a satisfies the hypothesis of (A). Let $h: L \to L'$ be the (g, g'^a)-equivariant $p.1.$ homeomorphism constructed in the last paragraph. Then by (29.2) $h_{\#}(g) = g'^a$. Hence, if $p > 2$, f is homotopic to the $p.1.$ homeomorphism h, by (29.6). When $p = 2$ there is, up to homotopy, exactly one homotopy-equivalence of each degree (an immediate consequence of (29.5)). The map $\lambda_1 z_1 + \lambda_2 z_2 + \ldots + \lambda_n z_n \to \lambda_1 \bar{z}_1 + \lambda_2 z_2 + \ldots + \lambda_n z_n$ induces a $p.1.$ homeomorphism of degree (-1) on $\mathbb{R}P^{2n-1}$. We leave it to the reader to find a $p.1.$ homeomorphism of degree $+1$. $\square$

Appendix

Chapman's Proof of the Topological Invariance of Whitehead Torsion

As this book was being prepared for print the topological invariance of Whitehead torsion (discussed in §25) was proved by Thomas Chapman[23]. In fact he proved an even stronger theorem, which we present in this appendix. Our presentation will be incomplete in that there are several results from infinite dimensional topology (Propositions A and B below) which will be used without proof.

Statement of the theorem

Let $I_j = [-1, 1], j = 1, 2, 3, \ldots,$ and denote

$$Q = \prod_{j=1}^{\infty} I_j = \text{the Hilbert cube}$$

$$I^k = \prod_{j=1}^{k} I_j$$

$$Q_{k+1} = \prod_{j=k+1}^{\infty} I_j.$$

It is an elementary fact that these spaces are contractible.

Main Theorem: *If X and Y are finite CW complexes then $f\colon X \to Y$ is a simple-homotopy equivalence if and only if $f \times 1_Q\colon X \times Q \to Y \times Q$ is homotopic to a homeomorphism of $X \times Q$ onto $Y \times Q$.*

Corollary 1 (Topological invarance of Whitehead torsion): *If $f\colon X \to Y$ is a homeomorphism (onto) then f is a simple-homotopy equivalence.*

PROOF: $f \times 1_Q\colon X \times Q \to Y \times Q$ is a homeomorphism. □

Corollary 2: *If X and Y are finite CW complexes then $X \underset{s}{\wedge} Y \Leftrightarrow X \times Q \approx Y \times Q$.*

PROOF: If $F\colon X \times Q \to Y \times Q$ is a homeomorphism, let f denote the composition $X \xrightarrow{\times 0} X \times Q \xrightarrow{F} Y \times Q \xrightarrow{\pi} Y$. Then $f \times 1_Q \simeq F$. Hence, by the

[23] His paper will appear in the American Journal of Mathematics. A proof not using infinite-dimensional topology of Corollary 1 for polyhedra has subsequently been given by R. D. Edwards (to appear).

Main Theorem, f is a simple homotopy equivalence. The other direction follows even more trivially. □

Results from infinite-dimensional topology

Proposition A: *If X and Y are finite CW complexes and $f: X \to Y$ is a simple-homotopy equivalence then $f \times 1_Q: X \times Q \to Y \times Q$ is homotopic to a homeomorphism of $X \times Q$ onto $Y \times Q$.*

COMMENT ON PROOF: This half of the Main Theorem is due to James E. West [*Mapping cylinders of Hilbert cube factors*, General Topology and its Applications 1, (1971), 111–125]. It comes directly (though not easily) from the geometric definition of simple-homotopy equivalence. For West proves that, if $g: A \to B$ is a map between finite CW complexes and $p: M_g \to B$ is the natural projection, then $p \times 1: M_g \times Q \to B \times Q$ is a uniform limit of homeomorphisms of $M_g \times Q$ onto $B \times Q$. This implies without difficulty that $p \times 1$ is homotopic to a homeomorphism. Recalling (proof of (4.1)) that an elementary collapse map be viewed as the projection of a mapping cylinder, it follows that if $f: X \to Y$ is a simple-homotopy equivalence (– a map homotopic to a sequence of elementary expansions and collapses) then $f \times 1: X \times Q \to Y \times Q$ is homotopic to a homeomorphism. □

Proposition B (Handle straightening theorem): *If M is a finite dimensional p.l. manifold (possibly with boundary) and if $\alpha: R^n \times Q \to M \times Q$ is an open embedding, with $n \geq 2$, then there is an integer $k > 0$ and a codimension-zero compact p.l. submanifold V of $M \times I^k$ and a homeomorphism $G: M \times Q \to M \times Q$ such that*

(i) $G|\alpha((R^n - \text{Int } B^n(2)) \times Q) = 1$, *($B^n(r)$ = ball of radius r)*

(ii) $G\alpha(B^n(1) \times Q) = V \times Q_{k+1}$,

(iii) *Bdy V (the topological boundary of V in $M \times I^k$, not its manifold boundary) is p.l. bicollared in $M \times I^k$.*

COMMENT: This theorem is due to Chapman [to appear in the Pacific Journal of Mathematics]. It is a (non-trivial) analogue of the Kirby-Siebenmann finite dimensional handle straightening theorem [$K-S$]. In the ensuing proof it will serve as "general position" theorem, allowing us to homotop a homeomorphism $h: K \to L$, K and L simplicial complexes, to a map (into a stable regular neighborhood of L—namely $M \times I^k$) which is nice enough that the Sum Theorem (23.1) applies.

Proof of the Main Theorem

In what follows X, Y, X', Y', will denote finite CW complexes unless otherwise stipulated.

Because Q is contractible, there is a covariant homotopy functor from the category of spaces with given factorizations of the form $X \times Q$ and maps between such spaces to the category of finite CW complexes and maps which is given by $X \times Q \mapsto X$ and $(F: X \times Q \to Y \times Q) \mapsto (F_0: X \to Y)$ where F_0 makes the following diagram commute

$$\begin{array}{ccc} X \times Q & \xrightarrow{F} & Y \times Q \\ \uparrow{\scriptstyle \times 0} & & \downarrow{\scriptstyle \pi} \\ X & \xrightarrow{F_0} & Y \end{array}$$

Explicitly, the correspondence $F \mapsto F_0$ satisfies

(1) $F \simeq G \Rightarrow F_0 \simeq G_0$

(2) $(GF)_0 \simeq G_0 F_0$

(3) If $f: X \to Y$ then $(f \times 1)_0 = f$.

[In particular $(1_{X \times Q})_0 = 1_X$.]

Definition:—*The ordered pair* $\langle X, Y \rangle$ *has Property P* iff $\tau(H_0) = 0$ *for every homeomorphism* $H: X \times Q \to Y \times Q$. *(The torsion of a non-cellular homotopy equivalence is defined following* (22.1).)

From Proposition A and from properties (1) and (3) above, the Main Theorem will follow once we know that every pair $\langle X, Y \rangle$ has Property P.

Lemma 1:—*If* $\langle X, Y \rangle$ *has Property P then* $\langle Y, X \rangle$ *has Property P.*

PROOF: If $H: Y \to X$ is a homeomorphism then so is $H^{-1}: X \to Y$ and, by assumption, $\tau((H^{-1})_0) = 0$. But $(H^{-1})_0$ is a homotopy inverse to H_0. Hence $\tau(H_0) = 0$, by (22.5). □

Lemma 2:—*If* $\langle X, Y \rangle$ *has Property P and if* $X \curlywedge X'$, $Y \curlywedge Y'$ *then* $\langle X', Y' \rangle$ *has Property P.*

Proof: Consider the special case where $Y = Y'$. Suppose that $H: X' \times Q \to Y \times Q$ is a homeomorphism. If $f: X \to X'$ is a simple-homotopy equivalence then, by Proposition A, there is a homeomorphism $F: X \times Q \to X' \times Q$ with $F \simeq f \times 1_Q$. Thus we have

$$X \times Q \xrightarrow{F} X' \times Q \xrightarrow{H} Y \times Q$$

$$X \xrightarrow{f} X' \xrightarrow{H_0} Y.$$

Since $\langle X, Y \rangle$ has Property P, $(HF)_0$ is a simple-homotopy equivalence. But $(HF)_0 \simeq (H_0 F_0) \simeq H_0 f$ where f is a simple-homotopy equivalence. Hence H_0 is a simple-homotopy equivalence. Therefore $\langle X', Y \rangle$ has Property P.

The general case now follows easily from the special case and Lemma 1. □

From this point on which shall introduce polyhedra into our discussion as though they were simplicial complexes (whereas in fact a polyhedron is a topological space X along with a family of piecewise-linearly related triangulations of X). This will in every case make sense and be permissible because of invariance under subdivision. (See (25.1) and (25.3).)

Lemma 3:—*If $\langle X, M\rangle$ has Property P whenever X is a simplicial complex and M is a p.*1. *manifold then all CW pairs have Property P.*

PROOF: If $\langle X', Y'\rangle$ is an arbitrary CW pair then (7.2) there are simplicial complexes X and Y'' such that $X' \wedge\!\!\!\!\searrow X$ and $Y' \wedge\!\!\!\!\searrow Y''$. Now let $j\colon Y'' \to R^N$ be a simplicial embedding into some large Euclidean space and let M be a regular neighborhood (see [Hudson]) of $j(Y'')$ in R^N. Then M is a $p.1.$ manifold and $M \searrow j(Y'')$. Hence $Y'' \xrightarrow{j} j(Y'') \subsetneq M$ is a simple-homotopy equivalence, so $Y' \wedge\!\!\!\!\searrow M$. Now lemma 3 follows from Lemma 2. □

In the light of Lemma 3, the Main Theorem will follow immediately from

Lemma 4:—*If X is a connected simplicial complex and M is a p.*1. *manifold then $\langle X, M\rangle$ has Property P.*

PROOF: The proof is by induction on the number r of simplexes of X which have dimension ≥ 2.

If $r = 0$ then dimension X is 0 or 1, in which case X has the homotopy type of a point or a wedge product of circles. Thus $\pi_1 X = \{1\}$ or $\pi_1 X = \mathbb{Z} * \mathbb{Z} * \ldots * \mathbb{Z}$, and, by (11.1) and (11.6), $Wh(X) = 0$. So Property P holds automatically

If $r > 0$ let σ be a top dimensional simplex of X; say $n = \dim \sigma \geq 2$. Let σ_0 be an n-simplex contained in Int σ such that $\sigma - \text{Int}\ \sigma_0 \overset{p.1.}{\cong} \dot{\sigma} \times I$. Denote $X_0 = X - \text{Int}\ \sigma_0$ and note that X_0 has a cell structure from $(X - \text{Int}\ \sigma) \cup (\dot{\sigma} \times I)$. The induction hypothesis applies to the complex $X - \text{Int}\ \sigma$, so $\langle X - \text{Int}\ \sigma, M'\rangle$ has Property P for any $p.1.$ manifold M'. Since $X_0 \searrow (X - \text{Int}\ \sigma)$, Lemma 2 implies that $\langle X_0, M'\rangle$ has Property P for all $p.1.$ manifolds M'.

Suppose that $H\colon X \times Q \to M \times Q$ is a homeomorphism. We must show that $\tau(H_0) = 0$.

Let $\beta\colon (R^n, B^n(1)) \to (\text{Int}\ \sigma, \sigma_0)$ be a homeomorphism. Then $\alpha = H \circ (\beta \times 1_Q)\colon R^n \times Q \to M \times Q$ is an open embedding with $n \geq 2$. Let k be a positive integer, V a $p.1.$ submanifold of $M \times I^k$, and $G\colon M \times Q \to M \times Q$ a homeomorphism satisfying the conclusion of Proposition B. Condition (i), that $G|\alpha((R^n - \text{Int}\ B^n(2)$ should be: $) \times Q) = 1$, implies $G \simeq 1_{M \times Q}$ since any homeomorphism of $B^n(2) \times Q$ onto itself which is the identity on $\dot{B}^n(2) \times Q$ is homotopic, rel $\dot{B}^n(2) \times Q$, to the identity on $B^n(2) \times Q$. (One simply views this as a homeo-morphism of

$$\left(B^n(2) \times \prod_{j=1}^{\infty} \left[\frac{-1}{2^j}, \frac{1}{2^j}\right]\right) \subset \ell_2 \text{ (Hilbert space)}$$

and takes the straight line homotopy). Thus $H \simeq GH$ and $H_0 \simeq (GH)_0$.

Let $\pi_k: M \times Q \to M \times I^k$ be the natural projection, and let $i: M \to M \times I^k$ be the zero section. We have the homotopy-commutative diagram

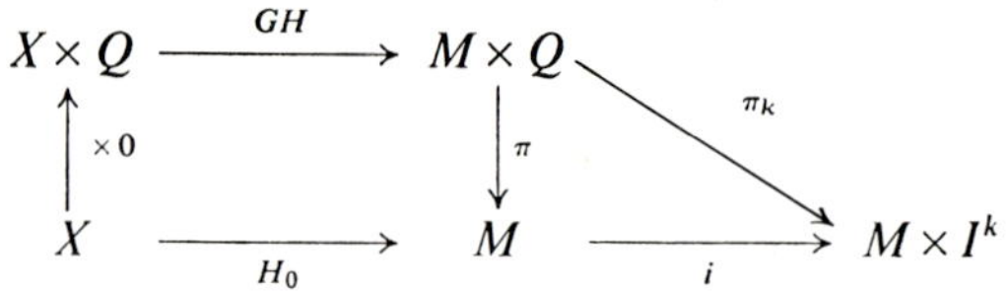

Denote $M_0 = \text{Closure}((M - I^k) - V)$. Conclusion (ii) of Proposition B implies that

(a) $GH(\sigma_0 \times Q) = V \times Q_{k+1}$
(b) $GH(Bdy\sigma_0 \times Q) = BdyV \times Q_{k+1}$,
(c) $GH(X_0 \times Q) = M_0 \times Q_{k+1}$.

Let $f = \pi_k GH(\times 0) \simeq iH_0$. Then (a)–(c) and the contractibility of Q and Q_{k+1} show that f restricts to homotopy equivalences $\sigma_0 \to V$, $Bdy\ \sigma_0 \to Bdy\ V$ and $X_0 \to M_0$. But $(M \times I^k, V, M_0)$ is a polyhedral triad (which can be triangulated as a simplicial triad), so we can use the Sum Theorem (23.1) to compute $\tau(f)$. Obviously $f|\sigma_0: \sigma_0 \to V$ and $f|Bdy\ \sigma_0: Bdy\ \sigma_0 \to Bdy\ V$ are simple-homotopy equivalences since the Whitehead groups involved vanish. To study $(f|X_0): X_0 \to M_0$ consider the commutative diagram

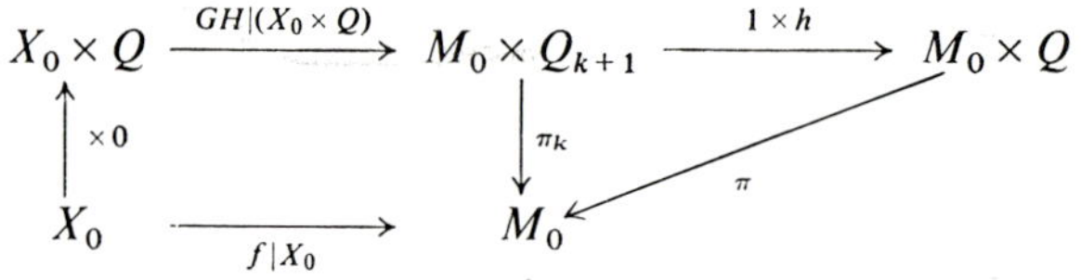

where $h: Q_{k+1} \to Q$ is the most obvious homeomorphism. Here M_0 is a $p.l.$ manifold since $Bdy\ V$ is $p.l.$ bicollared by conclusion (iii) of Proposition B. Hence $\langle X_0, M_0 \rangle$ has Property P. Thus $\tau(f|X_0) = 0$. Therefore, by the Sum Theorem, $\tau(f) = 0$.

But $f \simeq iH_0$ and $i: M \to M \times I^k$ is clearly a simple-homotopy equivalence. Hence H_0 is a simple-homotopy equivalence, as desired. □

Selected Symbols and Abbreviations

[ ] bibliographical reference

$\square$ the discussion of the proof is ended or omitted

$I = [0, 1]$

$I^r = I \times \ldots \times I$ (r-copies); $I^r \equiv I^r \times 0 \subset I^{r+1}$

∂ = boundary

J^r = closure of $\partial I^{r+1} - I^r$

$\searrow$ — strongly deformation retracts to

$\searrow$ — collapses to (in CW category) (§4)

$\nearrow$ — expands to (in CW category) (§4)

$\nearrow\!\!\searrow$ — expands and collapses to

$\oplus$ — direct sum of algebraic objects or disjoint union of spaces

M_f — the mapping cylinder of f

K^r — the r-skeleton of the complex K

$<$ — is a subcomplex of

$f_\#$ — the map between fundamental groups induced by f, or the map between groups of covering homeomorphisms of the universal covering spaces induced by f (depends on base points in either case) (§3)

$\theta(x, \tilde{x})$ — the isomorphism from fundamental group to the group of covering transformations of the universal cover (depends on choice of x and $\tilde{x}$) (§3)

$\mathbb{Z}$ — the integers

$\mathbb{Z}_p$ — the integers modulo p

$\mathbb{Z}(G)$ — the group ring of G

$Wh(L)$ — the Whitehead group of the complex L (§6; also §21)

$K_G(R)$ — an abelian group depending on ring the R and the subgroup G of its units (§10)

$Wh(G)$ — the Whitehead group of the group G (equals $K_T(\mathbb{Z}(G))$ where T = group of trivial units)

$Wh(\pi_1 L)$ — a group constructed by canonically identifying all the groups $Wh(\pi_1(L, x))$, L a connected complex (§19)

(R, G)-complex — a based chain complex over R with preferred class of bases (§12)

$Wh(G)$-complex — an (R, T)-complex, where $R = \mathbb{Z}(G)$ and T = group of trivial units of R

$\langle f \rangle_{x,y}$ — matrix of the homomorphism f with respect to bases x, y (§9)

$\langle x/y \rangle$ — matrix expressing elements of basis x in terms of elements of basis y (§9)

$f \oplus g : A \oplus B \to A' \oplus B'$ by $(f \oplus g)(a, b) = (f(a), g(b))$

$f+g:A \oplus B \to C$ by $(f+g)(a, b) = f(a)+g(b)$

$f+g:A \to B \oplus C$ by $(f+g)(a) = (f(a), g(b))$

$\simeq$ — homotopy of maps or homotopy equivalence of spaces

$\approx$ — homeomorphism of spaces

$\overset{s}{\sim}$ — stable equivalence of chain complexes (§14)

$\cong$ — isomorphism in the category under discussion

C_h — complex over R' induced by $h:R \to R'$ (§18)

ω_n — standard generator of $H_n(I^n, \partial I^n)$ (page 7)

$\langle \varphi_\alpha \rangle$ — equals $\varphi_{\alpha *}(\omega_n)$

Bibliography

J. W. Alexander

The combinatorial theory of complexes, Ann. of Math. **31**, (1930), 292–320.

D. R. Anderson

1. *The Whitehead torsion of the total space of a fiber bundle*, Topology **11** (1972), 179–194.
2. *Whitehead torsions vanish in many S^1 bundles*, Inventiones Math. **13**, (1971) 305–324.
3. *A note on the Whitehead torsion of a bundle modulo a subbundle*, Proc. AMS **32** (1972), 593–595.
4. *Generalized product theorems for torsion invariants with applications to flat bundles*, Bull. AMS **78** (1972), 465–469.

H. Bass

1. *Algebraic K-theory*, W. A. Benjamin, Inc., New York, 1968.
2. *K-theory and stable algebra*, Publ. de l'Inst. des Hautes Etudes Sci., #**22** (1964).

H. Bass, A. Heller, and R. Swan

The Whitehead group of a polynomial extension, Publ. de l'Inst. des Hautes Etudes Sci. #**22** (1964).

H. Bass, J. Milnor and J.-P. Serre

Solution of the congruence subgroup problem for SL_n ($n \geq 3$) and SP_{2n} ($n \geq 2$), Publ. Inst. des Hautes Etudes Sci. #**33** (1967).

E. M. Brown

The Hauptvermutung for 3-complexes, Trans. AMS, **144** (1969), 173–196.

C. Chevalley

Fundamental concepts of algebra, Academic Press, New York, 1956.

W. H. Cockroft and J. A. Combes

A note on the algebraic theory of simple equivalences, Quart. J. Math., **22** (1971), 13–22.

W. H. Cockroft and R. M. F. Moss

On the simple homotopy type of certain two-dimensional complexes, preprint.

M. Cohen

A general theory of relative regular neighborhoods, Trans AMS **136** (1969), 189–229.

M. N. Dyer and A. J. Sieradski

Trees of homotopy types of two-dimensional CW-*complexes*, preprint.

B. Eckmann and S. Maumary

Le groupe des types simple d'homotopie sur un polyèdre, Essays on Topology and Related Topics, Mémores dediés à Georges deRham, Springer-Verlag, New York, 1970, 173–187.

F. T. FARRELL and J. B. WAGONER

Algebraic torsion for infinite simple homotopy types, preprint.

G. HIGMAN

The units of group rings, Proc. London Math Soc. **46** (1940), 231–248.

P. J. HILTON and S. WYLIE

Homology theory, Cambridge U. Press, Cambridge, England, 1962.

H. HOSOKAWA

Generalized product and sum theorems for Whitehead torsion Proc. Japan Acad. **44** (1968) 910–914.

S. T. HU

Homotopy theory, Academic Press, New York (1959).

J. F. P. HUDSON

Piecewise linear topology, W. A. Benjamin, Inc. New York, 1968.

M. A. KERVAIRE, S. MAUMARY and G. DERHAM

Torsion et type simple d'homotopie, Lecture Notes in Mathematics, **48** (1967).

R. C. KIRBY and L. C. SIEBENMANN

On the triangulation of manifolds and the Hauptvermutung, Bull. AMS **75** (1969), 742–749.

K. W. KWUN and R. H. SZCZARBA

Product and sum theorems for Whitehead torsion, Ann. of Math. **82** (1965), 183–190.

J. MILNOR

1. *Whitehead Torsion*, Bull. AMS **72**, 1966, 358–426.
2. *Two complexes which are homeomorphic but combinatorially distinct*, Ann. of Math. **74** (1961), 575–590.

P. OLUM

1. *Self-equivalences of pseudo-projective planes*, Topology, **4** (1965), 109–127.
2. *Self-equivalences of pseudo-projective planes, II* Topology, **10** (1971), 257–260.

H. SCHUBERT

Topology, Allan and Bacon, Inc., 1968, Boston.

SEIFERT and THRELLFALL

Lehrbuch der topologie, Chelsea Publishing, New York N.Y.

L. C. SIEBENMANN

Infinite simple homotopy types, Indag. Math., **32** No. 5, 1970.

E. H. SPANIER

Algebraic topology, McGraw-Hill Book Co., New York, 1966.

J. STALLINGS

1. *Whitehead torsion of free products*, Annals of Math. **82**, (1965), pp. 354–363.

2. *On infinite processes leading to differentiability in the complement of a point*, Differential and Combinatorial Topology, (A Symposium in honor of M. Morse), Princeton University Press, Princeton, N.J., (1965) 245–254.

R. Stöcker

Whiteheadgruppe topologischer Räume, Inventiones Math. **9**, 271–278 (1970).

G. W. Whitehead

Homotopy theory, M.I.T. Press, Cambridge, Mass (1966).

J. H. C. Whitehead

1. *Simplicial spaces, nucleii and m-groups*, Proc. London Math. Soc. **45**, 1939, 243–327.
2. *On incidence matrices nucleii and homotopy types*, Ann. of Math. **42**, (1941) 1197–1239.
3. *Combinatorial homotopy*, I, Bull AMS, **55** (1949), 213–245.
4. *Simple homotopy types*, Amer. J. Math., **72**, 1952, pp. 1–57.

E. C. Zeeman

On the dunce hat, Topology, **2** (1964), 341–358.

Index

Graduate Texts in Mathematics

Soft and hard cover editions are available for each volume up to Vol. 14, hard cover only from Vol. 15.

1 Takeuti/Zaring. Introduction to Axiomatic Set Theory.
2 Oxtoby. Measure and Category. 2nd ed.
3 Schaeffer. Topological Vector Spaces.
4 Hilton/Stammbach. A Course in Homological Algebra.
5 MacLane. Categories for the Working Mathematician.
6 Hughes/Piper. Projective Planes.
7 Serre. A course in Arithmetic.
8 Takeuti/Zaring. Axiomatic Set Theory.
9 Humphreys. Introduction to Lie Algebras and Representation Theory. 2nd printing, revised.
10 Cohen. A Course in Simple Homotopy Theory.
11 Conway. Functions of One Complex Variable. 2nd ed.
12 Beals. Advanced Mathematical Analysis.
13 Anderson/Fuller. Rings and Categories of Modules.
14 Golubitsky/Guillemin. Stable Mappings and Their Singularities.
15 Berberian. Lectures in Functional Analysis and Operator Theory.
16 Winter. The Structure of Fields.
17 Rosenblatt. Random Processes. 2nd ed.
18 Halmos. Measure Theory.
19 Halmos. A Hilbert Space Problem Book.
20 Husemoller. Fibre Bundles. 2nd ed.
21 Humphreys. Linear Algebraic Groups.
22 Barnes/Mack. An Algebraic Introduction to Mathematical Logic.
23 Greub. Linear Algebra. 4th ed.
24 Holmes. Geometric Functional Analysis and Its Applications.
25 Hewitt/Stromberg. Real and Abstract Analysis. 4th printing.
26 Manes. Algebraic Theories.
27 Kelley. General Topology.
28 Zariski/Samuel. Commutative Algebra I.
29 Zariski/Samuel. Commutative Algebra II.
30 Jacobson. Lectures in Abstract Algebra I: Basic Concepts.
31 Jacobson. Lectures in Abstract Algebra II: Linear Algebra.
32 Jacobson. Lectures in Abstract Algebra III: Theory of Fields and Galois Theory.
33 Hirsch. Differential Topology.
34 Spitzer. Principles of Random Walk. 2nd ed.
35 Wermer. Banach Algebras and Several Complex Variables. 2nd ed.
36 Kelley/Namioka. Linear Topological Spaces.
37 Monk. Mathematical Logic.

38 GRAUERT/FRITZSCHE. Several Complex Variables.
39 ARVESON. An Invitation to C^*-Algebras.
40 KEMENY/SNELL/KNAPP. Denumerable Markov Chains. 2nd ed.
41 APOSTOL. Modular Functions and Dirichlet Series in Number Theory.
42 SEERE. Linear Representations of Finite Groups.
43 GILLMAN/JERISON. Rings of Continuous Functions.
44 KENDIG. Elementary Algebraic Geometry.
45 LOEVE. Probability Theory. 4th ed. Vol. 1.
46 LOEVE. Probability Theory. 4th ed. Vol. 2.
47 MOISE. Geometric Topology in Dimensions 2 and 3.
48 SACHS/WU. General Relativity for Mathematicians.
49 GRUENBERG/WEIR. Linear Geometry. 2nd ed.
50 EDWARDS. Fermat's Last Theorem.
51 KLINGENBERG. A Course in Differential Geometry.
52 HARTSHORNE. Algebraic Geometry.
53 MANIN. A Course in Mathematical Logic.
54 GRAVER/WATKINS. Combinatorics with Emphasis on the Theory of Graphs.
55 BROWN/PEARCY. Introduction to Operator Theory. Vol. 1: Elements of Functional Analysis.
56 MASSEY. Algebraic Topology: An Introduction.
57 CROWELL/FOX. Introduction to Knot Theory.
58 KOBLITZ. *p*-adic Numbers, *p*-adic Analysis, and Zeta-Functions.
59 LANG. Cyclotomic Fields.
60 ARNOLD. Mathematical Methods in Classical Mechanics.
61 WHITEHEAD. Elements of Homotopy Theory.
62 KARGAPOLOV/MERZLJAKOV. Fundamentals of the Theory of Groups.
63 BOLLOBAS. Graph Theory—An Introductory Course.
64 EDWARDS. Fourier Series, Volume 1. second edition.
65 WELLS. Differential Analysis on Complex Manifolds.
66 WATERHOUSE. Introduction to Affine Group Schemes.
67 SERRE. Local Fields.
68 WEIDMANN. Linear Operators in Hilbert Spaces.
69 LANG. Cyclotomic Fields II.
70 MASSEY. Singular Homology Theory.
71 FARKAS/KRA. Riemann Surfaces.
72 STILLWELL. Classical Topology and Combinatorial Group Theory.
73 HUNGERFORD. Algebra.
74 DAVENPORT. Multiplicative Number Theory.
75 HOCHSCHILD. Basic Theory of Algebraic Groups and Lie Algebras.
76 IITAKA. Algebraic Geometry.
77 HECKE. Lectures on the Theory of Algebraic Numbers.
78 BURRIS/SANKAPANAVAR. A Course in Universal Algebra.